M.T. Meeglio

Introduction to
Molecular Photochemistry

Chapman and Hall Chemistry Textbook Series

CONSULTING EDITORS

R. P. Bell, M.A., Hon. LLD., F.R.S., Professor of Chemistry at the University of Stirling

N. N. Greenwood, Ph.D., Sc.D., Professor of Inorganic and Structural Chemistry at the University of Leeds

R. O. C. Norman, M.A., D.Sc., Professor of Chemistry at the University of York

OTHER TITLES IN THE SERIES

Symmetry in Molecules J. M. Hollas

N.M.R. and Chemistry J. W. Akitt

Introduction to Molecular Photochemistry

C. H. J. Wells
Principal Lecturer in Chemistry
Kingston Polytechnic, Kingston upon Thames

LONDON
CHAPMAN AND HALL

First published 1972
by Chapman and Hall Ltd
11 New Fetter Lane, London EC4P 4EE

© 1972 C. H. J. Wells

Printed in Great Britain at The Pitman Press, Bath

SBN 412 11250 7

This limp bound edition is sold subject to the condition that it shall not, by way of trade or otherwise, be lent, re-sold, hired out, or otherwise circulated without the publisher's prior consent in any form of binding or cover other than that in which it is published and without a similar condition including this condition being imposed on the subsequent purchaser.

All rights reserved. No part of this book may be reprinted, or reproduced or utilized in any form or by any electronic, mechanical or other means, now known or hereafter invented, including photocopying and recording, or in any information storage and retrieval system, without permission in writing from the Publisher.

Published in the U.S.A.
by Halsted Press, a Division
of John Wiley & Sons, Inc.
New York

To PAM

Preface

The expansion of interest in photochemistry over the past decade is reflected by the large number of research papers and reviews of photochemical research which have been published during this period. The growth of photochemistry is connected to some extent to the realization by researchers that many of the concepts of molecular electronic spectroscopy are applicable to photochemistry, and a language linking photochemistry and electronic spectroscopy has now arisen. The research work over the past decade has laid the foundations for the understanding of photochemical and photophysical processes, and photochemistry is now being taught in most Universities, Polytechnics and Colleges as part of the undergraduate teaching programme. Although a number of texts on photochemistry have appeared in recent years these are mainly at final-year undergraduate and post-graduate level and it is felt that there is a need now for an introductory undergraduate text on the subject.

The object of the book is to present the principles of molecular photochemistry in a form suitable for undergraduate students new to photochemistry and to electronic spectroscopy. It is hoped that it may also be useful to post-graduate students and research workers seeking a concise introduction to photochemistry. The first chapter is concerned with the properties of electromagnetic radiation and the relationship between the absorption of radiation, absorption spectra and photochemistry. The second chapter develops the concepts of electronic transitions and absorption spectra, and this is followed in the third chapter by a discussion of electronically excited states and their properties. The fourth chapter deals with the kinetics of intramolecular and intermolecular processes, both chemical and physical, originating from excited state molecules. Some of the many photochemical reactions undergone by organic compounds and those of a limited number of inorganic compounds are outlined in the fifth and sixth chapters. The fifth chapter is concerned mainly with different types of photochemical reaction, while the sixth chapter deals mainly with photofragmentation and related reactions in various classes of compound.

I would like to express my appreciation to my colleagues, Mr E. F. H. Brittain and Dr A. Vincent, for reading the manuscript and for making valuable comments and suggestions.

C. H. J. Wells

Kingston upon Thames
January, 1972

Contents

	Preface	*page* ix
1	**Introduction**	
1.1	*Electromagnetic radiation*	1
1.2	*Photochemistry and absorption of radiation*	2
1.3	*Quantum yield*	8
1.4	*Experimental factors*	9
2	**Electronic transitions and spectra**	14
2.1	*Orbital types*	14
2.2	*Transition energies*	17
2.3	*Electronic states*	19
2.4	*Band positions*	21
2.5	*Intensities of bands*	22
2.6	*Selection rules*	24
2.7	*Spectra of diatomic molecules*	27
2.8	*Emission spectra*	30
3	**Electronically excited states**	34
3.1	*Intramolecular deactivation of excited states*	34
3.2	*Intermolecular deactivation of excited states*	48
4	**Kinetics of photochemical processes**	59
4.1	*Intramolecular processes*	59
4.2	*Intermolecular processes*	71

CONTENTS

5	Photochemical reactions – I	*page* 88
5.1	*Photoreduction*	88
5.2	*Photodimerization*	93
5.3	*Photo-addition*	97
5.4	*Photo-oxidation*	105
5.5	*Photorearrangement*	107
6	Photochemical reactions – II	116
6.1	*Bond dissociation*	117
6.2	*Photofragmentation and related reactions in various types of compound*	119
	Bibliography	140
	Index	141

Introduction 1

Photochemistry is the study of the physical processes or chemical changes which occur in molecules on absorption of ultraviolet-visible radiation. The sun is an extremely powerful source of radiation and consequently photochemical processes, such as photosynthesis, go on around us all the time. The early investigators studying the effect of light on chemical systems performed experiments by exposing the reaction systems to sunlight and attempting to elucidate the changes which had occurred after exposure for some period of time. Sunlight consists of a wide range of radiation and in the visible region alone it comprises the violet, blue, green, yellow, orange and red components. However in the early work no attempts were made to select any particular part of the sun's radiation with which to initiate the photochemical processes. Nowadays experiments are more carefully controlled and instead of using sunlight as the radiation source, researchers use lamps which emit narrow bands of electromagnetic radiation. Before going on to discuss the results of photochemical research it is necessary to consider some of the properties of electromagnetic radiation and the relationship between photochemistry and the absorption of radiation by molecular systems.

1.1 Electromagnetic radiation

Electromagnetic radiation can be considered in terms of electric and magnetic fields which oscillate sinusoidally in mutually perpendicular planes at right angles to the direction of propagation of the radiation. This situation is shown in Fig. 1.1 where, for clarity, plane polarized radiation is depicted.

INTRODUCTION TO MOLECULAR PHOTOCHEMISTRY

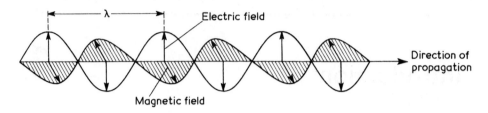

FIGURE 1.1
Plane polarized electromagnetic radiation

Electromagnetic radiation is described in terms of either the wavelength, λ, the frequency, ν, the wave number, $\bar{\nu}$, or the energy, E, of a quantum of radiation. The relationships between these terms and also the units used for these terms are summarized below:

Relationships	Units	Symbol
$\lambda = \dfrac{c}{\nu}$	λ, nanometre (10^{-9} metre)	nm
	ν, Hertz (cycles/second)	Hz
$\bar{\nu} = \dfrac{10^7}{\lambda}$	$\bar{\nu}$, reciprocal centimetre	cm^{-1}
$E = h\nu$	E, kilojoules per mole	kJ mol^{-1}

c is the speed of light = 3×10^8 m s^{-1}
h is Planck's constant = 6.626×10^{-34} J s

The wavelength of radiation changes throughout the electromagnetic spectrum (Fig. 1.2), and in the range of the ultraviolet-visible region, which is of main interest for photochemistry, the wavelengths vary from approx. 200 nm (2×10^{-7} m) to 700 nm (7×10^{-7} m). As can be seen from Fig. 1.2 this constitutes a very small range within the electromagnetic spectrum.

1.2 Photochemistry and absorption of radiation

Photochemical reactions occur when molecules absorb radiation and the resultant uptake of energy produces *excited* state molecules which can under-

INTRODUCTION

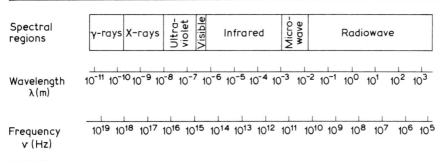

FIGURE 1.2
The electromagnetic spectrum

go chemical reaction. There are two principles in photochemistry relating to the absorption of radiation by molecules. One is the Grotthus-Draper principle which states that it is only the radiation which is absorbed that is photochemically active, and the other is the Stark-Einstein principle which states that a molecule can only absorb one quantum of radiation. These principles do not necessarily have complete applicability under all circumstances. For instance, with regard to the Stark-Einstein principle it has recently been shown that two-quantum absorption can be induced by the intense coherent radiation of laser beams. Nevertheless it can be taken that under the conditions of conventional photochemical experiments the Stark-Einstein principle will hold.

If the energy of the ground state molecule before the absorption of radiation is E_1 and that of the excited state molecule after absorption of radiation is E_2 then the energy, E, of the radiation required to bring about the transition is given by

$$E = E_2 - E_1 \qquad (1.1)$$

The expression for the wavelength of the radiation required for the transition can be derived from Equation (1.1) and the relationships given earlier connecting energy, frequency and wavelength. The expression is

$$\lambda = \frac{hc}{E_2 - E_1} \qquad (1.2)$$

For most organic molecules the energy difference between the ground state molecule and the excited state molecule of *lowest* energy is such that the wavelength of the absorbed radiation falls in the ultraviolet–visible region of the spectrum.

The absorption of radiation occurs in accordance with the Lambert and Beer laws. The Lambert law states that the fraction of incident radiation absorbed by a transparent medium is independent of the intensity of the radiation and that each successive layer of the medium absorbs an equal fraction of the incident radiation. The Beer law states that the amount of radiation absorbed is proportional to the number of molecules absorbing the radiation. That is, the amount of radiation absorbed is proportional to the concentration of the absorbing species. These laws can be combined to give the Beer–Lambert law which can be represented by

$$\log \frac{I_0}{I_T} \propto Cl \tag{1.3}$$

where I_0 and I_T are the intensities of the incident and transmitted radiation respectively, C is the concentration of the absorbing species, and l is the thickness of the absorbing layer. The term $\log I_0/I_T$ is called the *absorbance* and given the symbol, A. Using this symbol and introducing a proportionality constant into Equation (1.3) gives the Beer–Lambert law in the form in which it is commonly written:

$$A = \epsilon Cl \tag{1.4}$$

where ϵ, the proportionality constant, is referred to as the molar decadic absorption coefficient. The Beer–Lambert law is only strictly valid for radiation of a single wavelength (monochromatic radiation), and for each wavelength there will be a corresponding absorption coefficient. (The SI units for ϵ, c, and l in Equation (1.4) are $m^2\ mol^{-1}$, $mol\ dm^{-3}$ and mm respectively.)

The absorbance of a compound in solution can be measured using commercial spectrophotometers and the record of the variation of absorbance with change of wavelength of the incident radiation is its *absorption spectrum*. Since absorption coefficients can vary by as much as a factor of 10^5 it is often not practical to present an absorption spectrum as a plot of absorbance against wavelength. If, instead, $\log \epsilon$ is plotted against wavelength, peak maxima of widely differing ϵ-values can be shown on the one scale. This is illustrated for the absorption spectra of benzene and benzophenone in Figs. 1.3 (a) and (b) respectively.

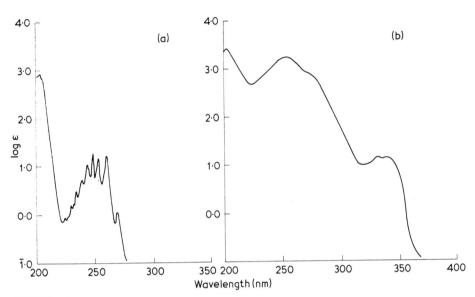

FIGURE 1.3
Absorption spectra of (a) benzene, and (b) benzophenone

Since the absorption of radiation results in the transition from a ground state of energy E_1 to an excited state of energy E_2 it might be expected that such a transition would be seen in the absorption spectrum as a single sharp line at a wavelength given by Equation (1.2). This is not the case, however, and a transition from a ground to an excited state is observed as an absorption band extending over a range of wavelengths with, in some cases, subsidiary peaks superimposed on the main band (see Fig. 1.3 (a)). The phenomena of band spectra as opposed to line spectra arise because molecules in any electronic state, ground or excited, possess vibrational and rotational energy in addition to electronic energy. The vibrational and rotational energy is quantized (i.e., confined to specific values) and for any electronic state there are a number of possible vibrational and rotational energy levels. This situation is represented schematically in Fig. 1.4. The vibrational energy levels are characterized by the vibrational quantum numbers $V = 0, 1 \ldots$ and $V' = 0, 1 \ldots$, whilst the rotational energy levels are characterized by the rotational quantum numbers $J = 0, 1 \ldots$ and $J' = 0, 1 \ldots$ The energy difference

INTRODUCTION TO MOLECULAR PHOTOCHEMISTRY

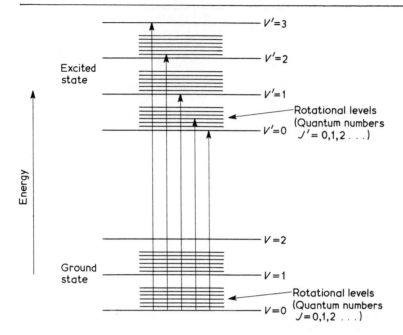

FIGURE 1.4
Vibrational and rotational energy levels in a ground state and in an excited state molecule

between the rotational levels is much less than that between the vibrational levels; typical rotational energy differences are 0.4 J mol^{-1} whereas typical vibrational energy differences are 4–400 J mol^{-1}.

At room temperature the majority of molecules in a system exist in the lowest vibrational level of the ground state, and hence transitions to the excited state normally originate from this level. Vibrational quantum numbers can change in going from the ground state to an excited state and therefore a number of possible transitions can originate from the $V = 0$ level of the ground state, e.g.,

$$V = 0 \rightarrow V' = 0, \; V = 0 \rightarrow V' = 1, \; V = 0 \rightarrow V' = 2 \text{ etc.}$$

Some of these possible transitions are represented by the vertical arrowed lines in Fig. 1.4. Since the different transitions correspond to the uptake of slightly different energies they will give rise to absorption peaks at different wavelengths. This feature can be seen in the band covering the range 220–

280 nm in Fig. 1.3 (a). If this band had been recorded using an instrument incapable of resolving the separate peaks a broad featureless band would have been observed. However, even using instruments of high resolving power it is not always possible to observe vibrational fine structure in an absorption band (see Fig. 1.3 (b)). Rotational fine structure is not normally observed in ultraviolet-visible absorption spectra because the energy difference between adjacent rotational energy levels is even smaller than that between vibrational levels.

So far the discussion has been confined to consideration of a single excited state of a molecule. There are, in fact, a number of possible excited states and a transition from the ground state to any one of these states will give rise to a band in the absorption spectrum. The spectrum shown in Fig. 1.3 (b) has two main bands centred at 252 and 342 nm. Each band represents a transition to a different excited state and the shorter the wavelength at which the band occurs the higher the energy of the excited state.

The energy required to form an excited state can be calculated from Equation (1.2) and the position of the peak maximum of the corresponding band in the absorption spectrum. For example, the longest wavelength band in the spectrum of benzophenone has a peak maximum at 342 nm, and the energy required to form the excited state corresponding to this band is

$$E_2 - E_1 = \frac{hc}{\lambda}$$

$$= \frac{6 \cdot 626 \times 10^{-34} \text{ (J s)} \times 2 \cdot 998 \times 10^8 \text{ (m s}^{-1})}{342 \times 10^{-9} \text{ (m)}}$$

$$= 5 \cdot 81 \times 10^{-19} \text{ J}$$

The above energy value is the energy required to produce one excited state molecule. The energy required to excite one mole of compound is called an *einstein*, and for the example under consideration the value is $6 \cdot 02 \times 10^{23} \times 5 \cdot 81 \times 10^{-19}$ J mol^{-1}, i.e. $3 \cdot 50 \times 10^5$ J mol^{-1}.

The energy of an einstein of radiation of wavelength λ can be calculated by substituting the numerical value of λ, when in units of nm, into the expression

$$\frac{1 \cdot 196 \times 10^8}{\lambda} \text{ J mol}^{-1} \qquad (1.5)$$

INTRODUCTION TO MOLECULAR PHOTOCHEMISTRY

Some values for the energy of radiation in the ultraviolet and visible regions are given in Table 1.1.

TABLE 1.1
Energy of ultraviolet-visible radiation.

Region	Approx. wavelength range (nm)	Wave number values $\bar{\nu}$ (cm^{-1})	Energy × 10^{-5} (J mol^{-1})
Ultraviolet	200 ↕ 400	50 000 25 000	5.95 2.98
Violet	↕ 450	22 222	2.65
Blue	↕ 500	20 000	2.39
Green	↕ 570	17 544	2.09
Yellow	↕ 590	16 949	2.03
Orange	↕ 620	16 129	1.93
Red	↕ 750	13 333	1.59

1.3 Quantum yield

According to the Stark-Einstein law the absorption of one quantum of radiation results in the formation of one photo-excited molecule. The photo-excited molecule may take part in a number of photochemical processes, either physical or chemical, and the quantum yield, Φ, for any one of these processes is defined by the expression

$$\Phi = \frac{\text{number of molecules undergoing the process}}{\text{number of quanta absorbed}}$$

An alternative form of this expression is

$$\Phi = \frac{\text{rate of process}}{\text{rate of absorption of radiation}}$$

INTRODUCTION

It is important when quoting a value for a quantum yield to specify the process to which the value refers. The reason for this can be seen by considering the photochemical reaction between hydrogen and chlorine. The mechanism for this reaction is as follows:

1. $Cl_2 + h\nu \rightarrow 2Cl$ photo-initiation

2. $Cl + H_2 \rightarrow HCl + H$ ⎫   chain
3. $H + Cl_2 \rightarrow HCl + Cl$ ⎭   propagation

4. $Cl \rightarrow \tfrac{1}{2}Cl_2$ (wall)   termination

The quantum yield for the primary process, viz., photo-initiation, cannot be greater than two since at maximum only 2Cl atoms can be produced per quantum absorbed. On the other hand, the quantum yield for the consumption of Cl_2 and H_2 is extremely large ($\Phi > 10^6$) because of the chain reaction occurring in the second and third steps. Thus in this reaction, as in many other reactions, there are a number of processes occurring each of which has a different quantum yield.

1.4 Experimental factors

Mercury discharge lamps are the most widely used sources of ultraviolet and visible radiation for conventional photochemical experiments. In this type of lamp a discharge is passed through mercury vapour and the electrically excited mercury atoms emit radiation. The relative intensity of the radiation at any wavelength is dependent upon the pressure of mercury inside the discharge tube. Typical spectral intensity distributions for lamps operating at low pressure ($\sim 10^{-6}$ bar) and medium pressure (~ 1 bar) are shown schematically in the bottom section of Fig. 1.5. It can be seen that the emission occurs at specific wavelengths and that as the pressure inside the lamp is increased so the emission occurs at a greater number of wavelengths. The emission from lamps operating at high pressure (~ 100 bar) is practically continuous over the ultraviolet and visible regions, and such lamps provide intense sources of radiation. The disadvantage with high pressure lamps is that they operate at

INTRODUCTION TO MOLECULAR PHOTOCHEMISTRY

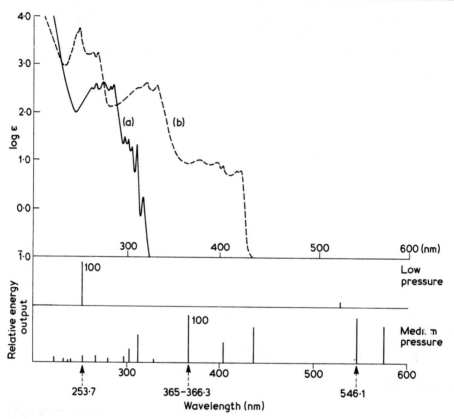

FIGURE 1.5
Top: absorption spectra of (a) naphthalene, and (b) anthraquinone. Bottom: typical relative outputs of low-pressure and medium pressure Hg-lamps

very high temperature so that water or forced air cooling of the surrounding quartz or glass envelope is necessary. Because of this and their relatively short lifetimes, high pressure lamps are not commonly used in commercial photochemical apparatus.

Since radiation has to be absorbed to be photochemically active, the choice of the source of radiation is governed by the absorption spectrum of the compound under study. The source chosen should have a high energy output within the wavelength range of the absorption band of interest in the reactant. For instance, benzene has a strong absorption in the region of

253·7 nm and because low pressure lamps generally emit about 90 per cent of their total radiation at this wavelength, low pressure lamps can be used to study the photoreactions of benzene.

It frequently happens in photochemical studies that it is necessary to excite selectively one component of a two component mixture. This can be achieved if the absorption spectra of the components are sufficiently different. For example, it can be seen from the spectra (a) and (b) in Fig. 1.5 that for a mixture of naphthalene and anthraquinone, the latter would absorb over the wavelength range 330–430 nm, whereas the former would not. Thus the anthraquinone could be excited selectively by irradiating the system with radiation in this range. This could be brought about by using a medium pressure mercury lamp in conjunction with a filter which absorbed all the lamp radiation below a wavelength of 330 nm.

Unwanted radiation can be 'cut-off' from the reaction system by the use of either glass or chemical filters. For the naphthalene-anthraquinone mixture, selective excitation of the anthraquinone could be achieved by inserting between the lamp and reaction vessel either of the filters (a) or (b) whose transmission characteristics are shown in Fig. 1.6.

Inspection of Fig. 1.5 indicates that the longest wavelength band in the spectrum of anthraquinone is probably overlapped at the short wavelength end by the relatively intense band centred at 320 nm and that this latter band could extend past 365–366·3 nm. Thus if it were required to excite selectively within the longest wavelength band of anthraquinone then a filter which cut-off the mercury radiation at 365–366·3 nm would have to be used. The filter combination (c) of Fig. 1.6 would be suitable. This filter is particularly useful if it is required to isolate radiation in the 400–408 nm region from the remainder of the mercury radiation.

Different types of glass have different transmission characteristics and this has to be borne in mind when setting up photochemical apparatus. Pyrex glass transmits down to about 310 nm whereas quartz glass will transmit down to about 200 nm. When radiation of wavelength less than 310 nm is to be used, any glass interface between the radiation source and the reaction system must be made of quartz.

Photochemical reactors may have the radiation source external to the system to be photolyzed or the source may be surrounded by a housing and immersed in the reaction mixture (Fig. 1.7). The former method is often the

INTRODUCTION TO MOLECULAR PHOTOCHEMISTRY

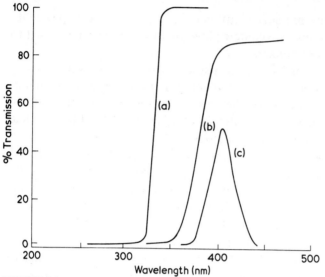

FIGURE 1.6
Radiation filters: (a) 2 mm Chance-Pilkington OW 12 filter, (b) 1 cm of naphthalene, 12·8 g/100 cm³ iso-octane, and (c) 10 cm of $CuSO_4 \cdot 5H_2O$, 0·44 g/100 cm³ 2·7M NH_4OH, plus 1 cm of I_2, 0·75 g/100 cm³ CCl_4, plus 1 cm of quinine hydrochloride, 2·00 g/100 cm³ H_2O.

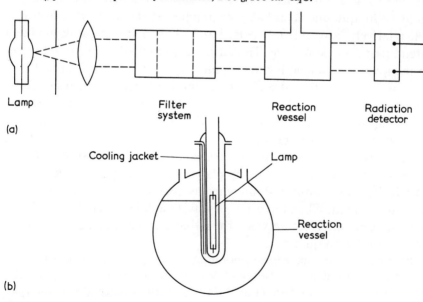

FIGURE 1.7
Experimental systems for (a) quantitative studies, and (b) preparative studies

INTRODUCTION

more convenient for quantitative studies where it is frequently necessary to use a combination of filters to isolate a narrow band of radiation. The latter method is commonly used in synthetic photochemistry where the use of a narrow band of radiation may not be so critical and where it is desirable to have as great a proportion as possible of the total lamp output incident upon the reactant mixture.

Electronic transitions and spectra

2

As indicated in the previous Chapter the absorption of ultraviolet or visible radiation by a molecule results in the molecule changing from the ground state to an excited state. The physical effect of the absorption of radiation is to cause the promotion of an electron from an orbital of lower energy to one of higher energy. The molecule with an electron in a higher energy orbital is then said to be in an excited state. Photochemistry is concerned with the physical and chemical reactions of excited state molecules. To gain an understanding of photochemical processes it is necessary to consider the types of molecular orbital involved in the initial electronic transition.

2.1 Orbital types

Molecular orbitals can be considered as being formed by the linear combination of the atomic orbitals of the atoms forming the molecular structure. The atomic and molecular orbitals represent electronic wave functions which relate to the probability of finding an electron in the region surrounding the nuclei, and they can be represented diagrammatically by enclosing the space around the nuclei in which there is a relatively high electron probability. Wave functions are mathematical expressions and can have either positive or negative values in the different regions of space around the nuclei (see Figs. 2.1 and 2.2). The combination of atomic orbitals leads to the formation of both bonding and antibonding molecular orbitals; the bonding orbitals having a high electron probability between the nuclei and the antibonding orbitals having a plane of zero electron probability between the nuclei. Bonding orbitals are of lower energy than the combining atomic orbitals while the antibonding orbitals are of higher energy.

2.1.1 Sigma orbitals

Sigma orbitals may be formed by combination of two one-electron atomic orbitals lying along the axis joining the atoms. The molecular orbitals resulting from the combination of 1s and 2p atomic orbitals are shown in Fig. 2.1. Bonding and antibonding sigma orbitals are designated by the symbols σ and σ^* respectively.

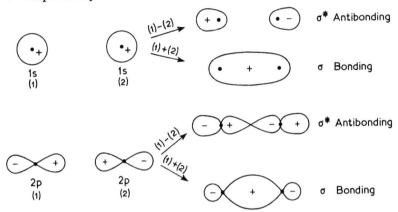

FIGURE 2.1
Sigma orbitals formed by linear combination of 1s and 2p atomic orbitals

The difference in energy between the bonding and antibonding orbitals is normally such that the promotion of an electron from a σ orbital to a σ^* orbital requires the absorption of radiation in the wavelength range 100–200 nm. In ethane, for example, the lowest energy $\sigma \rightarrow \sigma^*$ transition gives rise to an absorption band at 154 nm.

2.1.2 Pi orbitals

Pi orbitals may be formed by the sidewise overlap of atomic 2p orbitals. The overlap of two such orbitals will give a bonding π orbital and an antibonding π^* orbital having an approximate form as shown in Fig. 2.2.

The difference in energy between π and π^* orbitals is less than that between σ and σ^* orbitals with the result that $\pi \rightarrow \pi^*$ transitions give rise to absorption bands at longer wavelengths than $\sigma \rightarrow \sigma^*$ transitions. In the case of ethylene the $\pi \rightarrow \pi^*$ transition of lowest energy results in an absorption band at 200 nm.

There will be a number of π and π^* orbitals for a molecule possessing a delocalized π electron system. For example, there are six molecular orbitals in benzene formed by overlap of the six atomic 2p orbitals; three of these are π orbitals and three are π^* orbitals. The approximate form of these orbitals is shown in Fig. 2.3. The ultraviolet absorption spectrum of benzene (Fig. 2.4) exhibits three bands and each of these represents a transition of an electron from one of the π orbitals to one of the π^* orbitals.

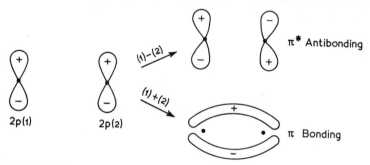

FIGURE 2.2
Pi orbitals formed by combination of two 2p atomic orbitals

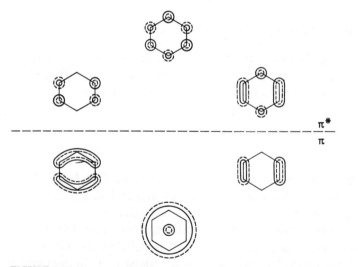

FIGURE 2.3
Representations of π bonding and π^* antibonding orbitals in benzene. (Dashed contours are for positive parts of the wavefunction, solid contours are for negative parts.)

ELECTRONIC TRANSITIONS AND SPECTRA

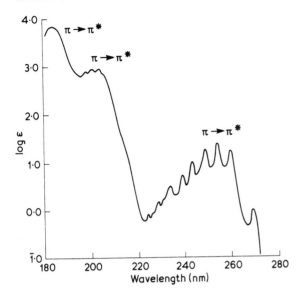

FIGURE 2.4
Absorption spectrum of benzene. (After Brittain, George and Wells, *Introduction to Molecular Spectroscopy*, Academic Press (1970).)

2.1.3 Non-bonding orbitals

Molecules containing heteroatoms have electrons in orbitals associated with the heteroatom which are not involved in the bonding system of the molecule. For instance, in carbonyl compounds there are two electrons in the non-bonding 2p orbital (n orbital) on the oxygen atom (Fig. 2.5). The absorption of radiation can lead to the promotion of one of these electrons into either a σ^* or a π^* orbital. Such transitions would be designated as n → σ^* or n → π^* transitions.

2.2 Transition energies

The energy required for an electronic transition is the difference between the energy of the orbital from which the electron originates and the orbital into which it is promoted. The relative energies of the different types of molecular

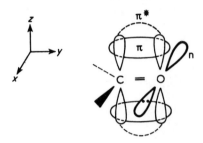

FIGURE 2.5
Representations of the n, π and π^* orbitals associated with the carbonyl group. The n orbital (2p) points along the x axis while the π and π^* orbitals lie in the yz plane

orbital are shown in Fig. 2.6. The possible transitions occurring between these orbitals are represented by the vertical arrowed lines. It can be seen that for a molecule containing σ, π and n orbitals the lowest energy transition is likely to be n → π^*. However it is possible that for molecules with a high degree of conjugation the π orbital can have higher energy than the n orbital and the $\pi \to \pi^*$ transition would then be the one of lowest energy.

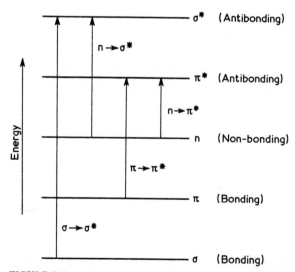

FIGURE 2.6
Molecular orbital energy levels and possible electronic transitions between orbitals. (After Higasi, Baba and Rembaum, *Quantum Organic Chemistry*, Interscience (1965))

ELECTRONIC TRANSITIONS AND SPECTRA

$\sigma \to \sigma^*$ and $n \to \sigma^*$ transitions are of relatively high energy and usually require the absorption of radiation of shorter wavelength than either $n \to \pi^*$ or $\pi \to \pi^*$ transitions. The wavelength of the radiation required for $\sigma \to \sigma^*$ and $n \to \sigma^*$ transitions is often shorter than 200 nm and hence inconvenient from a practical point of view. The majority of photochemical reactions studied have resulted from an initial $n \to \pi^*$ or $\pi \to \pi^*$ transition in the molecular system.

2.3 Electronic states

The multiplicity of an electronically excited state is defined by the expression $2S + 1$ where S is the algebraic sum of the spin quantum numbers of the electrons in the system. The spin quantum numbers can be either $+\frac{1}{2}$ or $-\frac{1}{2}$. As a consequence of the *Pauli exclusion principle* electrons in the same orbital must have their spins paired, i.e., one will have the spin quantum number $+\frac{1}{2}$ (represented by ↑) and the other $-\frac{1}{2}$ (represented by ↓). In molecules where all the electrons are paired in orbitals the sum, S, of the spin quantum numbers must equal zero and hence the multiplicity will be one. The molecule is then said to be in a *singlet* state; this state being designated by the symbol S. The ground states of the majority of organic molecules are singlet states.

If on promotion of an electron into a higher energy orbital the promoted electron retains its spin configuration the sum of the spin quantum numbers will still be equal to zero but the molecule will now be in an *excited* singlet state. If the spin configuration of the promoted electron is changed on excitation the spin of the promoted electron will not be paired with the spin of the electron in the vacated orbital. The spin quantum numbers of the unpaired electrons will be either $+\frac{1}{2} +\frac{1}{2}$ (↑↑) or $-\frac{1}{2} -\frac{1}{2}$ (↓↓). In such a case the algebraic sum of the spin quantum numbers will be one and the multiplicity of the state, $2S + 1$, will be three. Thus a molecule with two unpaired electrons is described as being in a *triplet* state. Triplet states are designed by the symbol T.

Molecules have minimum electronic energy when in the ground state, and when this is a singlet state it is given the symbol S_0. Promotion of an electron from the highest occupied molecular orbital (HOMO) in the ground state to the lowest vacant molecular orbital (LVMO) will require the minimum energy uptake consistent with a change in electronic configuration. The resultant state, which can be either singlet or triplet, will be the first excited state and is given the symbol S_1 or T_1. Singlet states are invariably of higher energy than the corresponding triplet states because of greater electron repulsion in the singlet state.

Excited states of greater energy than the S_1 and T_1 states may be formed by either exciting an electron from the HOMO orbital to an orbital of higher energy than the LVMO orbital or from an orbital of lower energy than the HOMO orbital to the LVMO orbital or one of higher energy. The higher energy singlet and triplet excited states are labelled $S_2, S_3 \ldots, T_2, T_3 \ldots$, in increasing order of energy.

The carbonyl group reacts photochemically in a large number of systems and consideration of the different types of singlet and triplet electronic states associated with this group illustrates concepts applicable to many other photochemically reactive species. In carbonyl compounds the filled orbitals of highest energy are often those associated with the carbonyl group and in many instances the relative energies of the σ, π and n orbitals of the carbonyl group are as shown in Fig. 2.6. This figure shows only the three filled orbitals of highest energy (σ, π and n) but obviously there will be other filled orbitals of lower energy in the molecular system. The latter orbitals are not shown because they are not normally involved in electronic transitions. The electrons in these lower energy orbitals are termed inner electrons.

The transition of lowest energy in the carbonyl compounds, which have an energy diagram as in Fig. 2.6, will be $n \rightarrow \pi^*$. Thus molecules in the first excited states (S_1 and T_1) will have n and π^* orbitals which contain a single electron. This situation can be indicated by using the full symbols $S_1(n, \pi^*)$ and $T_1(n, \pi^*)$ to represent the first excited states. An alternative designation is $^1(n, \pi^*)$ and $^3(n, \pi^*)$. The electronic configuration for molecules in these states and in the ground state S_0 can be written as below:

State	Electronic configuration	Electron spins			
		n	π^*	n	π^*
S_0	[Inner electrons] $\sigma^2 \pi^2 n^2$	↑↓		or ↓↑	
$S_1(n, \pi^*)$	[Inner electrons] $\sigma^2 \pi^2 n^1 (\pi^*)^1$	↑	↓ or ↓		↑
$T_1(n, \pi^*)$	[Inner electrons] $\sigma^2 \pi^2 n^1 (\pi^*)^1$	↑	↑ or ↓		↓

ELECTRONIC TRANSITIONS AND SPECTRA

One possible arrangement for the relative energies of the ground and the first three sets (S and T) of excited states of an aromatic carbonyl compound is given in Fig. 2.7. The type of transition leading to each of the excited states can be deduced from the symbols (n, π^*), (n, σ^*) and (π, π^*) alongside the states in the Figure.

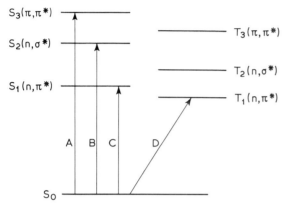

FIGURE 2.7
Typical energy level diagram for ground and excited states of a carbonyl compound

The absorption spectrum corresponding to the transitions A, B, C and D of Fig. 2.7 may appear as shown in Fig. 2.8. The intensities and positions of the bands in this spectrum are typical for aromatic carbonyl compounds but the assignment of the bands at 170 nm and 180 nm to $\pi \rightarrow \pi^*$ and n $\rightarrow \sigma^*$ transitions respectively is still open to question in a number of cases.

2.4 Band positions

The relative positions of bands in a spectrum can often be markedly affected by changing the solvent system. Changing from a non-polar solvent to a polar solvent shifts bands due to n $\rightarrow \pi^*$ transitions to shorter wavelengths and bands due to $\pi \rightarrow \pi^*$ transitions to longer wavelengths. Hydroxylic solvents are highly polar and when they are used as solvents for carbonyl compounds a hydrogen bond is formed between the oxygen atom of the carbonyl group and the hydrogen atom of the hydroxyl group. The hydrogen bonding is stronger for ground state molecules than for (n, π^*) excited state molecules where only one n electron on the oxygen atom is available for bonding. Thus

21

the energy of the ground state is lowered relative to that of the (n, π^*) excited state on changing to a hydroxylic solvent and the energy required for an n → π^* transition is increased. The shift to longer wavelength of the π → π^* band is thought to occur because the (π, π^*) excited state is more polar than the ground state and is stabilized to a greater extent than the ground state in

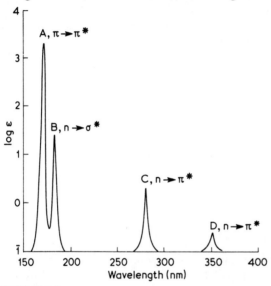

FIGURE 2.8
Representation of an absorption spectrum of an aromatic carbonyl compound. (Letters on the spectrum correspond to transitions given in Fig. 2.7)

polar solvents. The effect on the energy levels of singlet states on changing from a non-polar to a polar solvent is shown diagrammatically in Fig. 2.9. The energies of the triplet states will also alter on changing the solvent. Under circumstances where the S_1 and S_2 states and the T_1 and T_2 states are initially close in energy, changing to a polar solvent could interchange the energy positions of the (n, π^*) and (π, π^*) levels. This could then radically alter the photochemical behaviour of the system.

2.5 Intensities of bands

As can be seen from Fig. 2.8, there may be a wide range of intensities for the bands in an absorption spectrum. This arises because the electronic transitions corresponding to the absorption bands have different probabilities of occurrence. An intense band is associated with a transition of high probability and the transition is said to be *allowed*. A weak band is associated with a transition of low probability and the transition is said to be *forbidden*.

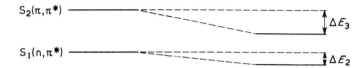

FIGURE 2.9
Alteration in energy levels of electronic states on changing from a non-polar to a polar solvent. ($\Delta E_3 > \Delta E_2 > \Delta E_1$)

One measure of the intensity of an absorption band is the value of the molar decadic absorption coefficient ϵ at the wavelength of maximum absorption. Another is the integrated absorption intensity. This is the molar decadic absorption coefficient integrated over the entire band and is given by

$$\text{integrated absorption intensity} = \int_{\bar{\nu}_1}^{\bar{\nu}_2} \epsilon \, d\bar{\nu} \qquad (2.1)$$

where $\bar{\nu}_1$ and $\bar{\nu}_2$ are the wave number limits of the absorption band.

The integrated absorption intensity is in turn related to the basic theoretical unit describing the intensity of an electronic transition, namely the oscillator strength f. The relationship is

$$f = 4.32 \times 10^{-9} \int_{\bar{\nu}_2}^{\bar{\nu}_2} \epsilon \, d\bar{\nu} \qquad (2.2)$$

INTRODUCTION TO MOLECULAR PHOTOCHEMISTRY

It is outside the scope of the present text to outline the theory leading to the derivation of Equation (2.2). The important point is that the theory shows that the value for the oscillator strength of a fully allowed transition is unity. By calculating the oscillator strength for any band from Equation (2.2) and comparing the value to unity it is often possible to decide whether the electronic transition giving rise to the band is allowed or forbidden. Typical values of the oscillator strengths of the bands in the absorption spectrum of an aromatic carbonyl compound are given in Table 2.1. The band with an oscillator strength of 0·5 represents an allowed transition whereas that with an oscillator strength of 0·02 represents a forbidden transition. The ϵ_{max} values for these bands are approximately 2000 and 20 m² mol⁻¹ respectively. It is not possible to give a definitive rule for deciding whether an ϵ_{max} value represents an allowed or forbidden transition. Frequently, however, bands with ϵ_{max} values above 1000 m² mol⁻¹ result from allowed transitions while those with ϵ_{max} values below 100 m² mol⁻¹ result from forbidden transitions.

TABLE 2.1
Typical oscillator strengths for transitions in aromatic compounds.

λ_{max}[a] (nm)	Transition	Oscillator strength, f	
170	$\pi \to \pi^*$	0.5	allowed
180	$n \to \sigma^*$	0.02	forbidden
280	$n \to \pi^*$	0.0004	forbidden
350	$n \to \pi^*$	10^{-5}	forbidden

[a] Wavelengths correspond to bands shown in Fig. 2.8.

2.6 Selection rules

The probability of occurrence of an electronic transition and hence the intensity of the associated absorption band is dependent upon various factors. These factors are included in *selection rules* which govern whether a transition will be allowed or forbidden. Transitions which conform to the selection rules can give rise to very intense absorption bands, i.e. bands with high f and high ϵ_{max} values. Transitions which do not conform to the selection

ELECTRONIC TRANSITIONS AND SPECTRA

rules either do not occur or else the probability of occurrence is so low that only very weak bands are observed in the spectrum.

The selection rules for polyatomic molecules can be summarized in relation to the oscillator strength f_a of a fully allowed $\pi \rightarrow \pi^*$ transition by the equation

$$f = p_s p_o p_p p_m f_a \qquad (2.3)$$

where f is the oscillator strength for the transition under consideration and the terms p_s, p_o, p_p and p_m are probability factors which respectively take into account the changes in electron spin, orbital symmetry, parity and momentum which occur as a result of the electronic transition. The probability factors are discussed below.

2.6.1 *Electron spin:* p_s

Selection rules predict that transitions which involve the change in spin of an electron during the excitation process are forbidden. A change in electron spin causes a change in the multiplicity and consequently transitions from singlet states to triplet states, or *vice versa*, are forbidden. Normally singlet to triplet (S → T) transitions are highly forbidden and the probability factor p_s can have values of the order of 10^{-5}. The selection rules break down when a heavy atom or a paramagnetic species is present in the system. When the spectrum is recorded under such conditions S → T transitions may be observed. The enhancing effects of iodine as a heavy atom and of oxygen as a paramagnetic molecule on singlet → triplet transitions are illustrated by the spectra in Fig. 2.10.

2.2.6 *Orbital symmetry:* p_o

The spatial properties of the orbitals involved in electronic transitions are of importance in determining the intensity of absorption bands. If the two orbitals involved in the transition do not simultaneously possess large amplitudes in the same region of space then the transitions are called overlap or space forbidden.

Inspection of Fig. 2.5 reveals that for the carbonyl group the π and π^* orbitals lie in the same plane and have a high degree of spatial overlap whereas the n and π^* orbitals lie in each other's nodal plane and have little spatial overlap. Thus $\pi \rightarrow \pi^*$ transitions in the carbonyl group are space allowed

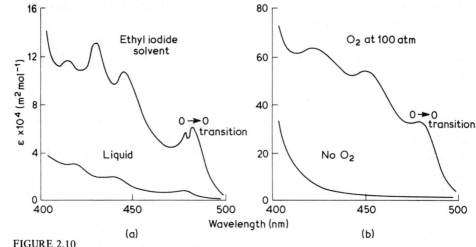

FIGURE 2.10
Enhancement of singlet → triplet transitions, (a) effect of ethyl iodide on spectrum of 1-chloronaphthalene, (b) effect of oxygen on spectrum of 2-naphthylphenylketone. The S → T bands show vibrational fine structure; the 0 → 0 transition is marked on each spectrum

while n → π^* transitions are space forbidden. The probability factor p_o for n → π^* transitions is generally about 10^{-2} in carbonyl compounds.

2.6.3 Parity: p_p

When the wave function of a molecule changes sign on reflection through a centre of symmetry it is termed ungerade (u), whereas if it does not change sign it is termed gerade (g). The bonding π and antibonding π^* orbitals in ethylene provide examples of wave functions which are ungerade and gerade respectively (Fig. 2.11).

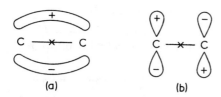

FIGURE 2.11
Representation of molecular orbitals in ethylene, (a) bonding π ungerade, (b) antibonding π^* gerade. (Note: the crosses on the diagrams represent the centres of symmetry.)

The parity selection rule states that electronic transitions from g → u and u → g are allowed but those from g → g and u → u are forbidden. In cases where parity 'forbiddeness' is of importance, such as for aromatic hydrocarbons, the probability factor p_p is about 10^{-1}.

2.6.4 *Momentum:* p_m

Any transition resulting in a large change in the linear or angular momentum of the molecule is momentum forbidden. For condensed ring compounds the probability factor p_m lies in the range 10^{-1} to 10^{-3}.

2.7 Spectra of diatomic molecules

It is instructive at this point to consider briefly the spectra of diatomic molecules since certain principles of absorption spectroscopy relevant to photochemistry can be discussed most easily in relation to diatomic spectra. Also, the spectra of diatomic molecules often show features which are not normally observable in the spectra of polyatomic molecules.

One aspect of diatomic molecules which favours discussion of their absorption spectra is the fact that the variation in potential energy with internuclear separation can be represented by a two-dimensional curve of the type shown in Fig. 2.12. Polyatomic molecules have polydimensional potential functions and these cannot be so readily represented on a two-dimensional page.

Each electronic state of a molecule will have a different potential energy curve and the relative positions of the curves for the ground state, S_0, and first excited states, S_1 and T_1, may be as shown in Fig. 2.12. The curves for the excited states will lie at higher energies than the ground state curve and will normally be displaced to greater internuclear separation than the ground state because of the weaker bonding in the energetically excited molecule.

The quantization of molecular vibrational energy is represented in Fig. 2.12 by the horizontal lines across the potential energy curves. At room temperature the majority of molecules are in the lower vibrational levels ($V = 0, 1, 2$) of the ground state and electronic transitions from the ground state can be considered as originating from these levels. There are no selection rules governing the change in vibrational level in going from one electronic state to another and transitions from the $V = 0$ level, say, of the S_0 state to any of the vibrational levels in the S_1 state are possible.

INTRODUCTION TO MOLECULAR PHOTOCHEMISTRY

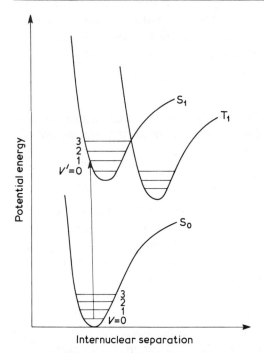

FIGURE 2.12
Typical plots of change in potential energy with internuclear separation for the ground state, S_0, and first excited states, S_1 and T_1, of a diatomic molecule

The time required for an electronic transition is very short ($\sim 10^{-15}$ s) compared to the time of a complete vibration of a diatomic molecule ($\sim 10^{-13}$ s). This time difference is taken into account in the Franck-Condon principle which states that the internuclear separation can be regarded as fixed during the time required for an electronic transition. Consequently electronic transitions can be represented on diagrams as in Fig. 2.12 by vertical lines connecting the initial and final states.

The relative intensities of the bands in the spectrum of a diatomic molecule are dependent upon the relative horizontal separations of the potential energy curves of the ground and excited state molecules. Three possible situations involving the ground and the first excited singlet state are given in Fig. 2.13 where in (a) the equilibrium internuclear separation (minimum of curve) is similar for both states, in (b) the equilibrium internuclear separation

28

is slightly greater in the excited state, and in (c) the equilibrium internuclear separation is much greater in the excited state.

The most probable transition in each situation of Fig. 2.13 is shown by the vertical arrowed line which starts at the point corresponding to the equilibrium internuclear separation in the ground state and terminates at the

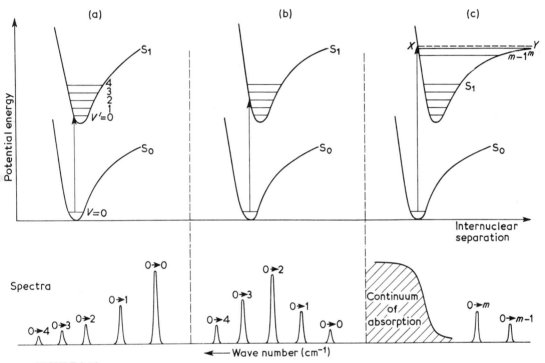

FIGURE 2.13
Relationship between relative positions of ground and excited state potential energy curves for a diatomic molecule and the absorption spectrum of the molecule

point where it meets the excited state curve at a vibrational level. In diagrams (a) and (b) the terminal points correspond to the $V' = 0$ and $V' = 2$ levels respectively and hence the $0 \to 0$ and $0 \to 2$ bands are the most intense in the spectra. In diagram (c) the transition finishes at a point X which lies above the dissociation level of the molecule and on vibrating from point X to the point Y the molecule will dissociate into the separate atoms. The spectrum will appear as shown in Fig. 2.13(c) and will have a continuum of absorption at

INTRODUCTION TO MOLECULAR PHOTOCHEMISTRY

energies corresponding to transitions from the ground state to points above the dotted horizontal line.

The schematic spectra of Fig. 2.13 suggest that the absorption spectra of diatomic molecules will exhibit bands corresponding to transitions to different vibrational levels of the excited state and that transitions from a particular level of the ground state ($V = 0$ in the cases shown) to successive vibrational levels in the excited state will give a series of bands in the spectrum. In fact, band series are commonly observed in the vapour phase spectra of diatomic molecules.

2.8 Emission spectra

The uptake of ultraviolet or visible radiation by a molecule results in the formation of an excited state or states of high energy. Such energy rich states are relatively short lived and rapidly lose the absorbed energy to return to the stable ground state. There are two types of intramolecular process whereby molecules can lose excess energy. One is classified as *radiative* since energy is lost by the emission of radiation, and the other as *non-radiative* since radiation is not emitted during energy loss. The former process is of importance in the present context since it gives rise to emission spectra. Both processes are important, however, in photochemistry since they represent the means whereby photochemically active excited states are degraded to the non-photochemically active ground state (see Chapter 3 for further discussion).

2.8.1 *Fluorescence and phosphorescence*

The radiation emitted during a radiative transition is termed *fluorescence* when the transition is between states of the same multiplicity and *phosphorescence* when the transition is between states of different multiplicity. That is, the emitted radiation is fluorescence for the $S_1 \rightarrow S_0$ transition and phosphorescence for the $T_1 \rightarrow S_0$ transition.

The relationship between the absorption spectrum of a molecule with a singlet ground state and its fluorescence and phosphorescence spectra is seen in Fig. 2.14. On absorption of radiation the molecule may be excited to an upper vibrational level of the first excited singlet state or end up in such a level after deactivation from an upper excited singlet state. The excess vibrational energy is rapidly lost by collisional deactivation and the molecule finishes up in the lowest vibrational level. Fluorescence arises from the radi-

ELECTRONIC TRANSITIONS AND SPECTRA

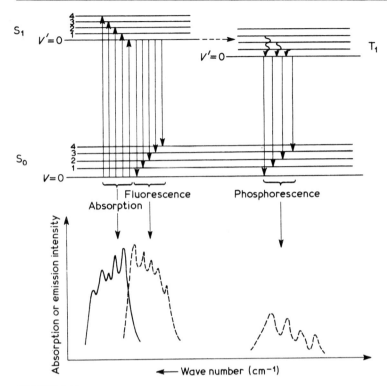

FIGURE 2.14
Origin of absorption, fluorescence and phosphorescence spectra. (Reproduced with permission from Brittain, George and Wells, *Introduction to Molecular Spectroscopy*, Academic Press (1970).)

ative transitions from the lowest vibrational level of the S_1 state to the various vibrational levels of the S_0 state. If the spacings of the vibrational levels in the excited state are similar to those in the ground state there will be an approximate 'mirror image' relationship between the absorption and fluorescence spectra. The spacing between the bands in the absorption spectrum is equal to the difference in energy between the vibrational levels of the excited state while that between the bands in the fluorescence spectrum is equal to the difference in energy between the vibrational levels of the ground state.

Although the 0 → 0 transitions in absorption and fluorescence are shown to have the same energy in the upper diagram of Fig. 2.14 this may not in

31

fact be the case, and in the absorption and fluorescence spectra (lower diagram) the bands arising from the 0 → 0 transitions may be slightly displaced. Consideration of Fig. 2.15 helps to explain why this can occur. If the excited state S_1 has a different solvation equilibrium from the ground state S_0 there will be a reorientation of the solvent cage after excitation and the more stable

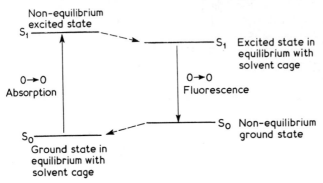

FIGURE 2.15
Effect of solute–solvent equilibrium on energy of electronic states

equilibrium state of S_1 will be formed. Fluorescence emission from this latter state will give a non-equilibrium ground state of higher energy than the original equilibrium ground state. Thus the 0 → 0 fluorescence transition will be of lower energy (lower wave number) than the original 0 → 0 absorption transition.

Phosphorescence results from the radiative transition from the lowest vibrational level of the T_1 state to the various vibrational levels of the S_0 state. Since the energy of the T_1 state is lower than that of the S_1 state the phosphorescence spectrum is observed at lower wave number values than the fluorescence spectrum. Since transitions between states of different multiplicity are forbidden the phosphorescence spectrum is weak. On the other hand, fluorescence is relatively intense since it corresponds to a transition between states of the same multiplicity.

The intensities of vibrational bands in fluorescence and phosphorescence spectra vary in a similar manner to bands in absorption spectra. Again this is most readily discussed in relation to diatomic molecules. Typical potential energy curves for the S_0 and S_1 states of a diatomic molecule are shown in Fig. 2.16. Here the excited state curve is displaced to greater internuclear

ELECTRONIC TRANSITIONS AND SPECTRA

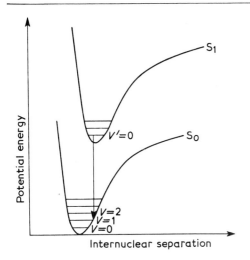

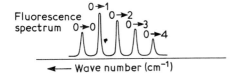

FIGURE 2.16
Representation of the fluorescence spectrum of a diatomic molecule showing possible relative intensities of the vibrational bands

separation than the ground state curve. The Franck–Condon principle applies to emission as well as to absorption, and emission processes can be represented by vertical lines connecting the initial and final states. In the situation represented in Fig. 2.16 the $0 \to 1$ transition is the most probable and the fluorescence spectrum may appear as shown.

33

Electronically excited states 3

The vast majority of photochemical reactions occur when a molecule is raised to either the first excited singlet state S_1 or the first triplet state T_1. Molecules in these states possess a large excess of energy and although this facilitates chemical reaction these highly energetic molecules are short lived and can often lose their excess energy and return to the stable ground state before reaction can occur. The likelihood of a photochemical reaction is dependent, therefore, upon the rates of the reaction processes as compared to the rates of the energy dissipation processes. The three types of process whereby the energy of electronically excited molecules can be dissipated are:

(1) Radiationless transitions from one electronic state to another (no radiation emitted during energy loss).
(2) Radiative transitions between electronic states (radiation emitted during energy loss).
(3) Electronic energy transfer between molecules.

In most cases, processes 1 and 2 are intramolecular and process 3 is intermolecular. The various intramolecular and intermolecular processes for the dissipation of electronic energy are discussed below.

3.1 Intramolecular deactivation of excited states

3.1.1 *Radiationless transitions: internal conversion and intersystem crossing*

Internal conversion (ic) is the term given to the radiationless process whereby a molecule transfers from one electronic state to another electronic state of the same multiplicity. Intersystem crossing (isc) is the term given to the process when the transfer involves electronic states of different multiplicity. These processes can be explained by consideration of Fig. 3.1. Here, for the sake of discussion, it has been assumed that the absorption of radiation promotes a molecule from the $V = 0$ level of its S_0 state to the $V' = 3$ level of its S_3 state. Once in this latter level the excess vibrational energy will be rapidly dissipated to the solvent medium (within a period of $\sim 10^{-13}$ s) and the

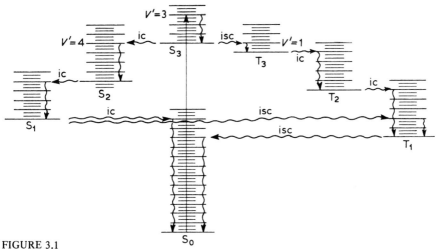

FIGURE 3.1
Possible radiationless deactivation processes for return of a molecule in the $V' = 3$ level of the S_3 state to the $V = 0$ level of the S_0 state. (Horizontal wavy lines represent internal conversion (ic) and intersystem crossing (isc). Vertical wavy lines represent vibrational deactivation by transfer of energy to the solvent.)

system will be deactivated to the $V' = 0$ level of the S_3 state. If there is any other electronic state whose potential energy function crosses that of the S_3 state and which has a vibrational level iso-energetic with the $V' = 0$ level of the S_3 state then, in principle, transfer to the other electronic state can occur. In the example of Fig. 3.1 the $V' = 4$ level of the S_2 state and the $V' = 1$ level of the T_3 state are shown as being iso-energetic with the $V' = 0$ level of the S_3 state and the internal conversion process ($S_3 \leadsto S_2$) and intersystem crossing process ($S_3 \leadsto T_3$) could occur. † These processes are followed by vibrational cascade to the $V' = 0$ levels of the S_2 and T_3 states and further

†Wavy arrows are used throughout the text to depict radiationless processes, and solid arrows are used to depict radiative processes. *Absorption* of radiation is depicted by a solid arrow.

internal conversions and intersystem crossings until the $V' = 0$ levels of the S_1 and T_1 states are reached. Once in these levels radiationless transitions to highly excited vibrational levels of the ground state can occur and these will be followed by a rapid drop to the $V = 0$ level of the ground state.

The rate of internal conversion between excited singlet states is extremely rapid (rate constants in the range 10^{11}–10^{13} s^{-1}) and consequently the lifetimes of upper excited singlet states are very short (10^{-11}–10^{-13} s) and but for a few exceptions deactivation to the S_1 state takes place before any radiative transition or photochemical reaction involving the upper state can take place. Similarly the lifetimes of upper triplet states are generally too short to allow radiative transitions or photoreactions to occur before deactivation to the T_1 state occurs. The most important intersystem crossing in reaching the T_1 state is the $S_1 \rightsquigarrow T_1$ process. Because of the spin interchange involved in this step the rate constant for the process is a factor of 10^2 to 10^6 less than that for internal conversions between upper states. Reverse intersystem crossing from the $V' = 0$ level of the T_1 state to the S_1 state is an endothermic process and cannot occur unless the triplet molecule can acquire the thermal energy to take it up to the vibrational level iso-energetic with the $V' = 0$ level of the S_1 state.

The rates of internal conversion and intersystem crossing processes are dependent upon the energy separation between the lowest vibrational levels of the states involved in the process; the larger the separation the slower the rate. The large energy separation between the lowest vibrational levels of the S_0 and S_1 states and between the lowest vibrational levels of the S_0 and T_1 states is partly responsible for the relative inefficiency of the $S_1 \rightsquigarrow S_0$ internal conversion and the $T_1 \rightsquigarrow S_0$ inter-system crossing. The rate constants for these processes are of the order of 10^6 to 10^{12} s^{-1} and 10^{-2} to 10^5 s^{-1} respectively.

3.1.2 Radiative transitions: fluorescence and phosphorescence

Once a molecule is in the $V' = 0$ level of the S_1 or T_1 states it can, as an alternative to internal conversion or intersystem crossing, return to the ground state by a radiative transition, i.e. by emitting radiation. The energy loss in dropping from the S_1 or T_1 state to the S_0 state is equal to the energy of the emitted radiation. Features of $S_1 \rightarrow S_0$ transitions (fluorescence) and of the $T_1 \rightarrow S_0$ transitions (phosphorescence) have been discussed in Chapter 2.

Emitted radiation can be detected experimentally and the duration of the fluorescence and phosphorescence emission from the S_1 and T_1 states of a molecule can often be measured. If the emission process is the only process depopulating the excited state then the time taken for the emission to decay to zero is the *natural radiative lifetime*, τ_0, of the excited state. However, other deactivation processes generally compete with the emission processes and since the lifetime of a state is determined by the sum of all the processes depopulating the state, the *observed radiative lifetime*, τ, is frequently less than the natural radiative lifetime.

When the absorption of radiation results in an allowed transition then the reverse transition giving rise to emitted radiation will also be allowed. In other words, states which are populated readily are also depopulated readily. The radiative lifetime of a state will be inversely proportional to the probability of populating that state and for states which are depopulated by allowed transitions the radiative lifetime will be short. The relationship between the radiative lifetime and the 'allowedness' of a transition is expressed approximately by the equation

$$\tau \cong \frac{10^{-5}}{\epsilon_{max}} \qquad (3.1)$$

where ϵ_{max} is the numerical value of the molar decadic absorption coefficient at the absorption band maximum.

Equation (3.1) can be used to estimate values for the radiative lifetimes of excited states. Allowed transitions from the S_0 to the S_1 state commonly have ϵ_{max} values in the region of 10^3 m² mol⁻¹, and in such cases the radiative lifetime of the S_1 state will be approximately 10^{-8} s. On the other hand, transitions from the S_0 to the T_1 state are spin forbidden and the ϵ_{max} values for such transitions can be of the order of 10^{-3} m² mol⁻¹. In such a case the radiative lifetime of the T_1 state would be approximately 10^{-2} s. In general, ϵ_{max} values for $S_0 \rightarrow S_1$ transitions in organic molecules are much greater than for $S_0 \rightarrow T_1$ transitions and consequently the radiative lifetimes of S_1 states are much shorter than those of T_1 states. That is, triplet state molecules will not decay to the ground state as rapidly as excited singlet state molecules and therefore triplet state molecules will exist in the system longer than excited singlet state molecules. This is of great significance for photochemistry since triplet state molecules being capable of existing for a relatively long time will have a high probability of colliding with other solute species and hence, other things being equal, will have a greater opportunity than singlet state molecules to take part in photochemical reactions.

Fluorescence can be classified as either *prompt* or *delayed* depending upon the period of time over which the fluorescence is emitted and the mechanism whereby the fluorescence arises. Prompt fluorescence is normally emitted within 10^{-6} s after the exciting radiation is extinguished and results from the S_1 state molecules, which are produced initially on excitation, reverting to the ground state. Delayed fluorescence is usually emitted over a much longer period of time than prompt fluorescence (typically $>10^{-3}$ s) and results from the radiative conversion to the ground state of S_1 state molecules which are formed by 'reaction' of the corresponding T_1 state molecules. Since delayed fluorescence depends upon the conversion of T_1 state species into S_1 state species this type of fluorescence can be emitted throughout the lifetime of the T_1 state of the molecule under study.

Prompt fluorescence can arise by radiative emission from either (a) an S_1 state molecule, or (b) an *excimer* formed by interaction of an S_1 state molecule with an S_0 state molecule. The term excimer is used to denote an excited state dimeric species. The two processes giving rise to prompt fluorescence are represented below:

(a) $S_1 \rightarrow S_0 + h\nu_f$

(b) $S_1 + S_0 \rightleftharpoons \underset{\text{excimer}}{(S_1 S_0)^*} \rightarrow 2S_0 + h\nu_{f(ex.)}$

The fluorescence emission of the excimer $(S_1 S_0)^*$ can differ from that of the 'free' S_1 state molecule, and the presence of excimers can sometimes be detected by recording the fluorescence spectrum with different concentrations of the solute present. As the concentration of the S_0 species increases so the probability of occurrence of process (b) above increases relative to that of process (a), and if there is a difference between the fluorescence spectrum at low and high concentration of the S_0 species then it is likely that excimers are present in the system. Excimers have been detected in solutions containing pyrene, where for dilute solutions ($\sim 10^{-6}$M) the fluorescence is violet and has a structured spectrum, while for more concentrated solutions ($\sim 10^{-3}$M) the fluorescence is blue and has a structureless spectrum.

The S_1 state molecules giving rise to delayed fluorescence can be formed by either (a) thermal excitation of a T_1 state molecule, or (b) collision of two

## ELECTRONICALLY EXCITED STATES

T_1 state molecules. The processes leading to delayed fluorescence are shown below:

(a) $T_1 \xrightarrow{\text{thermal excitation}} S_1 \rightarrow S_0 + h\nu_f$

(b) $T_1 + T_1 \rightarrow S_1 + S_0$
$S_1 \rightarrow S_0 + h\nu_f$

The first step in mechanism (b) above is referred to as energy pooling since two states combine their energy to form a third state of higher energy than either of the initial two. Normally the form of the delayed fluorescence spectrum is the same as that of the prompt fluorescence originating from the 'free' S_1 state molecule.

3.1.3 Radiative and radiationless transitions from S_1 and T_1 states

The absorption of radiation leads to excited state molecules which rapidly convert to the $V' = 0$ levels of the S_1 and T_1 states. If the excited state molecules do not participate in a chemical reaction they will all revert unchanged to the original ground state and the quantum yield for the re-formation of the ground state will be unity. This may be expressed as

$$\Phi = \frac{\text{no. of ground state molecules re-formed}}{\text{no. of ground state molecules initially excited}}$$

$$= 1 \cdot 00$$

The processes whereby molecules starting in the $V' = 0$ level of the S_1 state can return to the ground state are summarized in Fig. 3.2. Each of these processes will have a quantum yield defined by the expression:

$$\Phi_{\text{process}} = \frac{\text{no. of molecules deactivated by the process}}{\text{no. of ground state molecules initially excited}}$$

An alternative expression is:

$$\Phi_{\text{process}} = \frac{\text{rate of process}}{\text{rate of absorption of radiation}}$$

INTRODUCTION TO MOLECULAR PHOTOCHEMISTRY

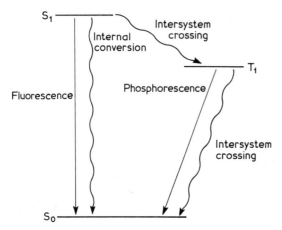

FIGURE 3.2
Intramolecular decay processes originating from S_1 and T_1 states

If there is no photochemical reaction and no energy loss from excited states by intermolecular quenching (see later) then the sum of the quantum yields for the deactivation processes originating from the S_1 state will equal unity. This can be written in the form

$$\Phi_f + \Phi_{ic} + \Phi_{isc}(S_1 \leadsto T_1) = 1 \cdot 00 \qquad (3.2)$$

where Φ_f, Φ_{ic} and Φ_{isc} are respectively the quantum yields for fluorescence, internal conversion and intersystem crossing.

When there is no photochemical reaction or energy loss by intermolecular quenching all the molecules which cross over to the T_1 state will decay either by emitting phosphorescence or by intersystem crossing to the S_0 state. The following expression will be applicable in these circumstances:

$$\Phi_{isc}(S_1 \leadsto T_1) = \Phi_p + \Phi_{isc}(T_1 \leadsto S_0) \qquad (3.3)$$

where Φ_p and Φ_{isc} are the quantum yields for phosphorescence and intersystem crossing.

Each decay process occurs at a rate which is represented by one of the unimolecular rate constants k_f, k_{ic}, $k_{isc}(S_1 \leadsto T_1)$, k_p and $k_{isc}(T_1 \leadsto S_0)$. The quantum yield for each process is related to these rate constants by the expression

$$\Phi_{process} = \frac{k_{process}}{\Sigma k_i} \tag{3.4}$$

where Σk_i is the sum of the rate constants for each of the processes.

The observed radiative lifetimes of the S_1 and T_1 states are determined by the rates of the deactivation processes originating from the state, and are given by the equation

$$\tau = \frac{1}{\Sigma k_i} \tag{3.5}$$

The extent to which each of the processes of internal conversion, intersystem crossing, fluorescence and phosphorescence contribute to the return of excited state molecules to the ground state varies for different molecular systems, and within the one system varies for different environmental conditions. Whether the deactivation of excited state molecules occurs primarily from the S_1 state or from the T_1 state depends on a number of factors, some of which are outlined below. These factors affect the rate of the $S_1 \leadsto T_1$ intersystem crossing relative to the rates of fluorescence and $S_1 \leadsto S_0$ internal conversion; if k_{isc} is large compared to k_f and k_{ic} then deactivation from the T_1 state will be the main route to the ground state. One factor which determines the relative magnitudes of k_{isc}, k_f and k_{ic} is the energy gap, $\Delta E(S_1 - T_1)$, between the $V' = 0$ levels of the S_1 and T_1 states. The results given in Table 3.1

TABLE 3.1
Energy separation ΔE between S_1 and T_1 states and quantum yields of processes originating from these states for some aromatic compounds in alcohol-ether glass at 77°K.[a]

Compound	$\Delta E(S_1 - T_1)$ (cm^{-1})	Φ_f	Φ_{isc} ($S_1 \leadsto T_1$)	Φ_p	Φ_p/Φ_f	$\Phi_f + \Phi_{isc}$ ($S_1 \leadsto T_1$)
Naphthalene	10 500	0.29	0.82	0.03	0.09	1.11
1-methylnaphthalene	10 450	0.43	0.47	0.02	0.05	1.00
Quinoline	10 200	0.05	0.96	0.10	1.9	1.01
Phenanthrene	7 200	0.12	0.86	0.13	1.1	0.98
Benzaldehyde	1 800	0.00	0.99	0.49	10^3	0.99
Benzophenone	1 750	0.00	1.00	0.74	10^3	1.00

[a] Data adapted from V. L. ERMOLAEV, *Soviet Phys., Usp.*, 333 (Nov.-Dec. 1963).

show that as ΔE decreases so the ratio Φ_p/Φ_f increases. This indicates that as ΔE decreases so deactivation via the T_1 state assumes greater importance.

The results of Table 3.1 are for molecules held in a rigid glass of alcohol-ether at 77° K and the observation that the sum of $\Phi_f + \Phi_{isc}(S_1 \leadsto T_1)$ is close to unity in each case suggests that the $S_1 \leadsto S_0$ internal conversion does not occur under these conditions. Phase and temperature apparently play a role in determining the relative rates of the processes since results obtained for the same molecules in fluid solution at room temperature indicate that the $S_1 \leadsto S_0$ internal conversion competes effectively with the other processes as a deactivation route.

The rate of internal conversion from the S_1 state to the S_0 state is dependent upon the energy gap, $\Delta E(S_1 - S_0)$, between the lowest vibrational levels of the S_1 and S_0 states; the smaller the energy gap the greater the rate of internal conversion. Similarly the rate of the $T_1 \leadsto S_0$ intersystem crossing is dependent upon the energy difference between the lowest vibrational levels of the T_1 and S_0 states. As the energy difference decreases so the rate of the $T_1 \leadsto S_0$ intersystem crossing increases relative to the rate of phosphorescence. This is illustrated by the values shown in Table 3.2.

The rate of fluorescence can be enhanced relative to the other processes deactivating the S_1 state by changing the temperature of the system. The

TABLE 3.2
Energy separation between T_1 and S_0 states and rate constants for $T_1 \leadsto S_0$ intersystem crossing and for phosphorescence in some aromatic compounds in alcohol-ether glass at 77°K.[a]

Compound	$\Delta E(T_1 - S_0)$ (cm^{-1})	$k'_{isc}(T_1 \leadsto S_0)$ (s^{-1})	k_p (s^{-1})	k'_{isc}/k_p
Acetophenone	25 750	1.7×10^2	2.8×10^2	0.61
Benzaldehyde	24 950	3.5×10^2	3.4×10^2	1.03
Carbazole	24 600	6.3×10^{-2}	6.9×10^{-2}	0.91
Triphenylamine	24 500	5.7×10^{-1}	8.6×10^{-1}	0.66
Phenanthrene	21 700	2.6×10^{-1}	4.6×10^{-2}	5.65
1-methylnaphthalene	21 000	4.5×10^{-1}	2.0×10^{-2}	22.5

[a] Data adapted from V. L. ERMOLAEV, *Soviet Phys., Usp.*, 333 (Nov.-Dec. 1963); and J. G. CALVERT and J. N. PITTS, Jr., *Photochemistry* (Wiley, New York, 1966).

fluorescence quantum yield, Φ_f, varies with temperature according to the equation:

$$\log\left(\frac{1}{\Phi_f} - A\right) = \frac{B}{T} \tag{3.6}$$

where B is a constant, A is equal to $(1 + \Sigma k'_i \tau)$ with τ being the observed radiative lifetime of the S_1 state and $\Sigma k'_i$ the sum of the rate constants for deactivation of the S_1 state by processes other than fluorescence. It can be seen from Equation (3.6) that the fluorescence yield can be increased by decreasing the temperature of the system.

TABLE 3.3
Variation of ϕ_f, ϕ_p, and τ_p for napthalene and τ_p for 1-chloronapthalene in heavy atom solvents.

Solvent	Naphthalene			1-Chloronaphthalene
	ϕ_f	ϕ_p	$\tau_p(s)$	$\tau_p(s)$
EPA [a]	0.55	0.05	2.5	0.23
Propyl chloride	0.44	0.08	0.52	0.075
Propyl bromide	0.13	0.24	0.14	0.06
Propyl iodide	0.03	0.35	0.076	0.023

[a] Mixture of ether, isopentane and ethanol.

The rate of the $S_1 \leadsto T_1$ intersystem crossing can be enhanced relative to the rates of fluorescence and $S_1 \leadsto S_0$ internal conversion by substituting a heavy atom into the molecule or by using solvents containing heavy atoms. The increased rate for the $S_1 \leadsto T_1$ intersystem crossing is shown up by the higher values for Φ_p and the lower values for Φ_f in the systems with heavy atoms present. The presence of heavy atoms also increases the rate of the $T_1 \leadsto S_0$ intersystem crossing and reduces the phosphorescence lifetime of the T_1 state. The results presented in Table 3.3 are indicative of these effects.

Another factor which influences the relative rates of the various deactivation processes is the nature of the S_1 and T_1 states; the relative rates being dependent upon whether the excited state is (n, π^*) or (π, π^*). This is considered further in the following section.

3.1.4 *Properties of* (n, π^*) *and* (π, π^*) *states*

The majority of reactions in organic photochemistry involve unsaturated compounds and, apart from a few exceptions, these reactions involve either the S_1 or the T_1 state of the molecule. For unsaturated compounds these states are normally either (n, π^*) or (π, π^*). The photophysical properties of these states determine to a large extent the photochemistry of molecular species and in many cases it is the difference in the properties of (n, π^*) and (π, π^*) states which gives rise to differences in photochemical behaviour. The properties of (n, π^*) and (π, π^*) states may be considered in relation to the S_1 and T_1 states of benzophenone which are (n, π^*) and the S_1 and T_1 states of naphthalene which are (π, π^*). These are particularly suitable model compounds for discussion since their photophysical and photochemical properties have been extensively investigated and they are representative of a large number of photoreactive compounds.

The values for the quantum yields of fluorescence, $S_1 \rightsquigarrow T_1$ intersystem crossing and phosphorescence in benzophenone and naphthalene in rigid solvent at 77° K are set out below:

	S_1 state	T_1 state	Φ_f	Φ_{isc} $(S_1 \rightsquigarrow T_1)$	Φ_p
Benzophenone	(n, π^*)	(n, π^*)	0.00	1.00†	0.74
Naphthalene	(π, π^*)	(π, π^*)	0.29	0.81	0.03

† Result for fluid solution at room temperature

Inspection of these figures shows that the $S_1 \rightsquigarrow T_1$ intersystem crossing between (n, π^*) states is highly favoured and that the large population of the T_1 state consequent upon this is shown up by the high value for the phosphorescence quantum yield. Corroboration for the favoured $S_1 \rightsquigarrow T_1$ intersystem crossing between (n, π^*) states is given by the immeasurably low quantum yield for fluorescence. In comparison to the 100 per cent transfer from the S_1 to the T_1 state in benzophenone only about 81 per cent of the naphthalene molecules in the S_1 state cross over to the T_1 state. The value for the quantum of fluorescence in naphthalene shows that the remainder return to the S_0 state by the emission of fluorescence rather than by $S_1 \rightsquigarrow S_0$ internal conversion.

There are two factors contributing to the high probability of $S_1 \rightsquigarrow T_1$ intersystem crossing between (n, π^*) states as compared to that between (π, π^*) states:

(1) The initial n → π^* excitation is a partly forbidden transition and hence the reverse transition will also be partly forbidden. Thus the lifetimes of S_1 (n, π^*) states tend to be greater than S_1 (π, π^*) states and the probability of conversion to the T_1 state correspondingly higher for S_1 (n, π^*) states.
(2) The energy separation between S_1 and T_1 states of the (n, π^*) type is normally small, within the range 1500–5000 cm^{-1}, whereas the energy separation for (π, π^*) states is frequently much larger, 10 000–15 000 cm^{-1}.

Another significant difference between the properties of the (n, π^*) states of benzophenone and the (π, π^*) states of naphthalene is that the phosphorescence quantum yields are markedly different. Whereas 74 per cent of the benzophenone molecules populating the T_1 state phosphoresce only about 3 per cent of the naphthalene molecules do so. If $T_1 \rightsquigarrow S_0$ intersystem crossing is the only other mechanism for deactivating the T_1 state then this process must be much more effective for the T_1 (π, π^*) state of naphthalene. This can be accounted for by the difference in the energy separations ΔE ($T_1 - S_0$) for the two types of T_1 state; ΔE generally being smaller when T_1 is (π, π^*) than when T_1 is (n, π^*).

A knowledge of quantum yield values is of importance in that these values can give information on the likely photochemical behaviour of molecules. For example, the high quantum yield value for intersystem crossing in benzophenone indicates that for aromatic ketones with S_1 and T_1 states of the (n, π^*) type photochemical reaction could probably only be initiated from the T_1 state. On the other hand, for aromatic hydrocarbons with S_1 and T_1 states of the (π, π^*) type photochemical reaction could probably occur from either the S_1 or the T_1 state.

It should be emphasized that the lowest energy triplet state need not be of the same type as the lowest energy excited singlet state. For example, the lowest energy triplet state of 2-acetonaphthone is (π, π^*) while the lowest energy excited singlet state is (n, π^*). Such a situation often arises when the energy separation between S_1 and S_2 states is small. The energy separation between the S_1 and S_2 states in benzophenone is larger than that in 2-acetonaphthone and here the lowest energy triplet state is of the same type as the S_1 state, i.e. both are (n, π*).

The relative energies of the first two sets of excited states (S and T) of 2-acetonaphthone and benzophenone are shown in Fig. 3.3(a) and (b) respectively. The T_2 (n, π*) state of 2-acetonaphthone is of lower energy than the

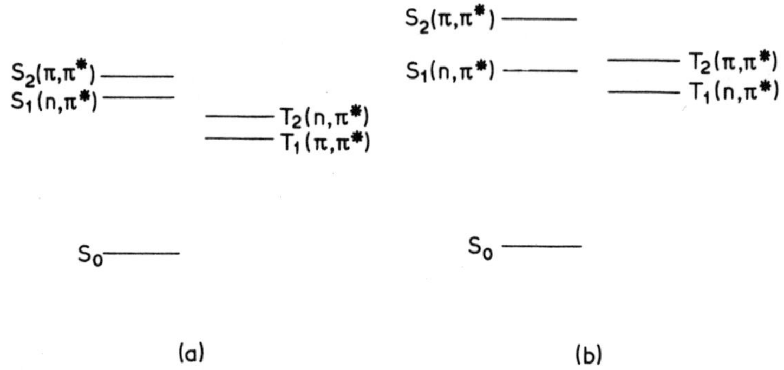

```
                                 S₂(π,π*) ─────
 S₂(π,π*) ─────                  S₁(n,π*) ─────              ───── T₂(π,π*)
 S₁(n,π*) ─────                                               ───── T₁(n,π*)
                     ───── T₂(n,π*)
                     ───── T₁(π,π*)


   S₀ ─────                           S₀ ─────

       (a)                               (b)
```

FIGURE 3.3
Relative energies of singlet and triplet states in (a) 2-acetonaphthone, and (b) benzophenone

S_1 (n, π*) state and therefore any intersystem crossing from the S_1 (n, π*) state will take place to the T_2 (n, π*) state in preference to the T_1 (π, π*) state. Population of the T_1 (π, π*) state will then occur by a rapid $T_2 \leadsto T_1$ internal conversion step. The mechanism for the population of the T_1 (n, π*) state of benzophenone differs from that for the T_1 state of 2-acetonaphthone in that the former state is formed by direct intersystem crossing from the S_1 (n, π*) state.

2-acetonaphthone and benzophenone are known to react photochemically via their lowest triplet states and as stated above the lowest triplet state of 2-acetonaphthone is (π, π*) whereas that for benzophenone is (n, π*). This difference in the nature of the reactive triplet states partly accounts for the difference in the photochemical behaviour of these two compounds (page 89).

It is obviously important in photochemical studies to determine the nature and the energies of the lowest excited singlet and triplet states of photochemically reactive molecules. Fortunately this information can be obtained sometimes from either the absorption or emission spectra of the molecule. For instance, the energy of the lowest vibrational level of the S_1 state can be estimated from the wave number values of the $0 \rightarrow 0$ vibrational bands of the $S_0 \rightarrow S_1$ and $S_1 \rightarrow S_0$ transitions in the absorption and fluorescence spectra respectively. The energy of the lowest vibrational level of the T_1 state can be obtained similarly from the wave number values of the $0 \rightarrow 0$ vibrational bands for the $S_0 \rightarrow T_1$ and $T_1 \rightarrow S_0$ transitions in the absorption and phosphorescence spectra respectively. It has to be borne in mind, however, that it is not always possible to locate the $0 \rightarrow 0$ vibrational band in absorption and emission spectra and, also, that many molecules do not give emission spectra.

Information as to whether the S_1 state of a molecule is (n, π^*) or (π, π^*) can often be obtained from the intensity of the longest wavelength band in the absorption spectrum of the molecule. For example, the longest wavelength band in the spectrum of biacetyl is of low intensity ($\epsilon_{max} = 1.8$ m^2 mol^{-1}) and it may be concluded from this that the band represents an n $\rightarrow \pi^*$ transition and that the S_1 state is an (n, π^*) state. Confirmation that the band corresponds to an n $\rightarrow \pi^*$ transition could be obtained by changing the polarity of the solvent and noting any change in the position of the band (page 21) and in the vibrational fine structure of the band. In non-polar solvents bands due to n $\rightarrow \pi^*$ transitions normally have well-defined vibrational fine structure which broadens on changing to a polar solvent. If the longest wavelength band in the absorption spectrum has an ϵ_{max} value greater than 100 m^2 mol^{-1} the lowest energy transition is probably $\pi \rightarrow \pi^*$ and the S_1 state is probably a (π, π^*) state. Again solvent studies could help to confirm this assignment.

Information on the nature of the T_1 state of a molecule can sometimes be obtained by recording the absorption spectrum of the molecule under a high pressure of oxygen. The presence of oxygen enhances $S_0 \rightarrow T_1$ transitions and since the increase in intensity of the band corresponding to the $S_0 \rightarrow T_1$ transition is much greater when the T_1 state is (π, π^*) than when it is (n, π^*) it may be possible to decide from the observed enhancement whether the T_1 state is (π, π^*) or (n, π^*).

Since $S_0 \rightarrow S_1$ bands arising from $\pi \rightarrow \pi^*$ transitions are intense relative to $S_0 \rightarrow S_1$ bands arising from n $\rightarrow \pi^*$ transitions it is to be expected on the

INTRODUCTION TO MOLECULAR PHOTOCHEMISTRY

basis of Equation (3.1) that the lifetimes of S_1 states of the (π, π^*) type will be shorter than those of S_1 states of the (n, π^*) type. Hence measurement of the fluorescence lifetimes of the S_1 state should enable the type of state to be assigned. The radiative lifetimes of molecules with T_1 states of the (π, π^*) type are normally of the order of 0.1–100 s while for molecules with states of the (n, π^*) type the lifetimes are normally of the order of 10^{-3}–10^{-2} s.

Some of the criteria used for identifying S_1 and T_1 states are listed in Table 3.4.

TABLE 3.4
Identifying features of S_1 and T_1 states of the (n, π^*) and (π, π^*) type.

	(n, π^*)	(π, π^*)
Absorption spectra		
ϵ_{max} values	$S_0 \rightarrow S_1$, $\epsilon_{max} < 10$	$S_0 \rightarrow S_1$, $\epsilon_{max} > 100$
O_2 perturbation	$S_0 \rightarrow T_1$, not normally enhanced	$S_0 \rightarrow T_1$, normally enhanced
$\Delta E(S_1 - T_1)$	frequently in range 1500–5000 cm^{-1}	frequently in range 10 000–20 000 cm^{-1}
solvent polarity	$n \rightarrow \pi^*$ bands shifts to shorter λ, vibrational fine structure broadens on increasing solvent polarity	$\pi - \pi^*$ bands shift to longer λ, vibrational fine structure unchanged on increasing solvent polarity
Emission spectra		
fluorescence	$\tau_f > 10^{-6}$ s $\phi_f < 0.01$	$\tau_f \sim 10^{-9} - 10^{-7}$ s $\phi_f \sim 0.5 - 0.05$
phosphorescence	$\tau_p \sim 10^{-3} - 10^{-2}$ s $\phi_p \sim 0.5 - 0.05$	$\tau_p \sim 0.1 - 100$ s $\phi_p \sim 0.5 - 0.05$

3.2 Intermolecular deactivation of excited states

Excited state molecules can lose their excess energy and return to the ground state by bimolecular processes involving other molecular species in the system. These processes are additional to the intramolecular processes described earlier so that the overall rate of decay of an excited state is the sum of the decay rates for the intramolecular and *inter*molecular routes. An

ELECTRONICALLY EXCITED STATES

example of an equation representing the rate of decay of a molecule in the T_1 state is given below:

$$\frac{-d[T_1]}{dt} = k_p [T_1] + k'_{isc} [T_1] + k_r [T_1]$$

$$+ k_1 [T_1][R] + k_2 [T_1][M] + k_3 [T_1]^2 + k_4 [T_1][A]$$

where k_p, k'_{isc} and k_r are respectively the rate constants for phosphorescence, $T_1 \leadsto S_0$ intersystem crossing and intramolecular reaction of the excited state molecule, and k_1, k_2, k_3, k_4 are respectively the rate constants for the following bimolecular processes:

(1) Quenching of the T_1 state by chemical reaction with reactant molecule R.
(2) Deactivation of the T_1 state by collision with substrate molecule M.
(3) Self-quenching of T_1 states by bimolecular collision.
(4) Deactivation of the T_1 state by donation of electronic energy to an acceptor molecule A.

For the purposes of the discussion the ground state and lowest triplet state of the photo-excited molecule are represented by the symbols D (S_0) and D*(T_1) respectively. The bimolecular processes given above involving the lowest triplet state species can then be written in chemical equation form as shown below:

(1) Chemical reaction   D*(T_1) + R → products
(2) Collisional deactivation   D*(T_1) + M → D(S_0) + M
(3) Self-quenching   D*(T_1) + D*(T_1) → D(S_0) + D(S_0)
(4) Energy transfer   D*(T_1) + A → D(S_0) + A*

In the chemical reaction depicted by process (1) the energy of the triplet state molecule is sufficient to overcome the activation energy barrier to the reaction. An example of a chemical reaction involving a triplet state species is seen in the reaction of triplet state benzophenone with isopropanol (see also page 89):

$$(C_6H_5)_2C=O^*(T_1) + (CH_3)_2CHOH \rightarrow (C_6H_5)_2\dot{C}OH + (CH_3)_2\dot{C}OH$$

Process (2) is a normal collisional deactivation where on collision the

INTRODUCTION TO MOLECULAR PHOTOCHEMISTRY

energy of the triplet state molecule is given up as heat to the surroundings. The rate for such a collision process cannot be greater than the rate at which the molecules diffuse through the solvent medium. Hence the upper limit for the rate constant will be equal to the rate constant for a diffusion controlled bimolecular process. An approximate value for the latter rate constant can be calculated from the modified Debye equation given below:

$$k_{\text{diff}} = \frac{8RT}{2\eta} \times 10^3 \text{ dm}^3\text{mol}^{-1}\text{s}^{-1} \tag{3.7}$$

where R is the gas constant (8.3 JK^{-1}mol^{-1}), T is the temperature (°K), and η is the solvent viscosity (N s m^{-2}). Typically for organic solvents at room temperature the value of k_{diff} is within the range $1 \times 10^{10} - 4 \times 10^{10}$ dm^3 mol^{-1} s^{-1}.

The bimolecular collision of triplet state molecules as represented by process (3) is a comparatively rare occurrence because of the short lifetimes of excited state species and because of their low concentration in reaction systems. When such collisions occur they can give rise to the emission of delayed fluorescence (see page 38).

The transfer of electronic excitation energy from a triplet state donor molecule to an acceptor molecule as shown in process (4) is a common occurrence in photochemical systems and represents an efficient method for the deactivation of photo-excited molecules. Minute traces of impurity can deactivate potentially photoreactive molecules by this process and for this reason it is important in photochemical work to rigorously purify both solutes and solvents. Oxygen is a particularly effective quencher of triplet state molecules ($k_4 \simeq 10^9 - 10^{10}$ dm^3 mol^{-1} s^{-1}) and for reactions involving triplet states it is imperative to degas the system to remove dissolved oxygen.

There are a number of mechanisms for the transfer of electronic energy from a donor molecule to an acceptor molecule; the different mechanisms being operative under different conditions. The principal mechanisms are described by the terms radiative transfer, short-range transfer and long-range transfer. Each of these transfer process are discussed in the following.

3.2.1 Radiative transfer

Radiative transfer of energy from a donor molecule D* to an acceptor molecule A involves the emission of radiation by the photo-excited donor, when in either the S_1 or T_1 state, followed by the absorption of the emitted radiation by the acceptor. The generalized mechanism for radiative transfer can be written as:

Donor emission $D^* \rightarrow D + h\nu$
Acceptor absorption $A + h\nu \rightarrow A^*$

The efficiency of the transfer is dependent upon the degree of similarity between the positions and intensities of the bands in the emission spectrum of the donor and the absorption spectrum of the acceptor; the greater the similarity the higher the percentage of the emitted radiation which will be re-absorbed.

3.2.2 Short-range transfer

Energy transfer from a donor D^* to an acceptor A can occur if the intermolecular separation between the two species approaches the collision diameter. It is not actually necessary for the two species to collide since transfer can occur at distances slightly greater than the collision diameter.

One restriction on the transfer of energy by the short-range process is that the multiplicities of the states involved in the transfer must be in accord with the *Wigner spin conservation rule*. The states of interest in molecular photochemistry are the S_0, S_1 and T_1 states and the transfer reactions of importance which involve these states and which obey the conservation rule are:

$$D^*(S_1) + A(S_0) \rightarrow D(S_0) + A^*(S_1) \qquad (3.8)$$

$$D^*(T_1) + A(S_0) \rightarrow D(S_0) + A^*(T_1) \qquad (3.9)$$

In the reaction given in Equation (3.8) an excited singlet state donor D^* transfers its energy to a ground singlet state acceptor A with the resultant formation of a ground singlet state donor D and an excited singlet state acceptor A^*. The formation of this latter species can often be detected by examining the fluorescence emission of the system whence the fluorescence of A^* is seen and not that of D^*. In the reaction shown in Equation (3.9) a triplet state donor molecule transfers its energy to a ground state acceptor molecule to give a ground singlet state donor molecule and a triplet state acceptor molecule. The observed emission in this system would be the phosphorescence of A^*.

The energy transfer reactions depicted in Equations (3.8) and (3.9) are referred to as singlet–singlet energy transfer and triplet–triplet energy transfer respectively. The efficiency of the singlet–singlet and triplet–triplet transfer processes depends on the relative energies of the states involved in the transfer step. If the energy of the excited state of the donor is greater than that of the excited state of the acceptor the transfer is relatively efficient but when the reverse is the case the transfer is inefficient.

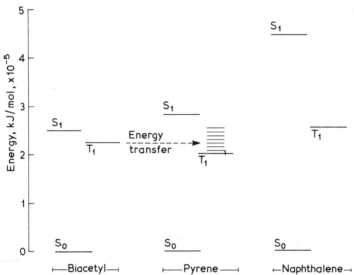

FIGURE 3.4
Relative energies of the S_1 and T_1 states of biacetyl, pyrene and naphthalene

The relative energies of the S_1 and T_1 states of biacetyl, pyrene and naphthalene are as shown in Fig. 3.4. Biacetyl absorbs radiation of longer wavelength than either naphthalene or pyrene (its S_1 state is of lower energy) and hence biacetyl can be selectively excited to its S_1 state in the presence of either pyrene or naphthalene. Energy transfer from the S_1 state of biacetyl to either the S_1 state of pyrene or the S_1 state of naphthalene is inefficient because this requires an uptake of energy. The T_1 state of biacetyl will be populated by intersystem crossing from its S_1 state and if pyrene is present in the system then transfer will occur from the $V' = 0$ level of the biacetyl T_1 state to an iso-energetic vibrational level of the pyrene T_1 state (see Fig. 3.4). The excess vibrational energy in the pyrene T_1 state will be lost rapidly and a pyrene molecule in the $V' = 0$ level of the T_1 state formed. Since the $V' = 0$ level of the naphthalene T_1 state is of higher energy than the $V' = 0$ level of the pyrene T_1 state triplet–triplet energy transfer will be inefficient in this case.

Singlet–singlet and triplet–triplet energy transfer is of importance in photochemistry since the tranfer represents a process whereby potentially reactive species can be deactivated or potentially reactive species can be formed. Triplet–triplet energy transfer can be of particular importance in the latter respect since in molecules where the $S_1 \rightsquigarrow T_1$ intersystem crossing is inefficient the energy transfer process can provide a route for enhancing the population of the T_1 state or, indeed, may be the only route for populating the T_1 state. The benzophenone–naphthalene system provides an example where triplet–triplet energy transfer enhances the population of a T_1 state. Selective excitation of benzophenone in the presence of naphthalene leads to the emission of naphthalene phosphorescence by the route shown in Fig. 3.5. The quantum yield of naphthalene phosphorescence is representative of the yield of the T_1 state of naphthalene, and for the benzophenone-naphthalene

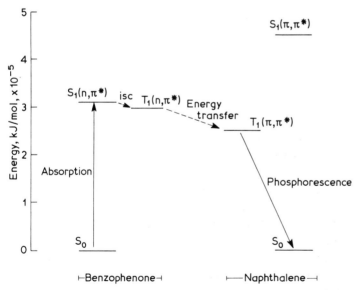

FIGURE 3.5
Mechanism of the benzophenone sensitized emission of naphthalene phosphorescence

system in benzene at 29° the quantum yield has a value of 0.07. The corresponding value when naphthalene is irradiated under the same conditions is 0.03. The higher quantum yield in the former case arises because of the greater efficiency of the $S_1 \rightsquigarrow T_1$ intersystem crossing in benzophenone ($\Phi_{isc} = 0.99$) as compared with that in naphthalene ($\Phi_{isc} = 0.39$). The energy transfer from benzophenone to naphthalene is diffusion controlled and once benzophenone triplet molecules are formed their energy is rapidly transferred to naphthalene to give naphthalene triplet state molecules.

Energy transfer between the benzophenone moiety and the naphthalene moiety can also occur by the short-range mechanism when the two groups are in the *same* molecule. Such *intra*molecular energy transfer has been shown to occur in molecules of the following type:

$$\text{Ph-CO-C}_6\text{H}_4\text{-(CH}_2)_n\text{-Naphthyl} \qquad n = 1, 2 \text{ or } 3$$

The absorption spectrum of a molecule of this type is virtually a composite of the absorption spectra of 4-methyl-benzophenone and 1-methyl-naphthalene. It is possible to selectively excite the benzophenone group in the 'combined' molecule and it is found in such circumstances that the emitted phosphorescence corresponds to that of the naphthalene group. This observation can be explained in terms of the relative energies of the different electronic states in the 'combined' molecule. The disposition of electronic energy levels in molecules of this type will be similar to that for the separate molecular entities, i.e. the relative positions of the energy levels will be as given in Fig. 3.5. Absorption of radiation within the lowest energy band in the spectrum of the 'combined' molecule will result in the excitation energy being initially localized in the benzophenone $-C=O$ chromophore, in which case the S_1 state will be (n, π^*). Intersystem crossing to the corresponding (n, π^*) triplet state will occur rapidly and this will be followed by *intramolecular energy transfer* to the (π, π^*) triplet state associated with the naphthalene group. Phosphorescence emission characteristic of a (π, π^*) state will then be emitted from the 'naphthalene' triplet state.

Singlet–singlet energy transfer from the naphthalene moiety to the benzophenone moiety can be induced in the 'combined' molecule by selective excitation of the naphthalene group. As can be seen from Fig. 3.5 the relative positions of the energy levels of the S_1 states favours such a transfer.

3.2.3 *Long-range transfer*

In certain situations transfer of electronic energy from a donor to an acceptor can occur when the donor and acceptor molecules are separated by a distance much greater than the donor–acceptor collision diameter. This can happen in cases where the energy differences between the vibrational levels of the ground and first excited states of the donor correspond to the energy differences between the vibrational levels of the ground and first excited states of the acceptor.

In the situation shown in Fig. 3.6 the differences in energy between the donor vibrational levels connected by the lines labelled A, B and C corres-

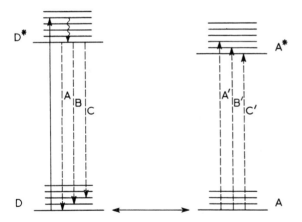

FIGURE 3.6
Long-range energy transfer from an excited state donor, D*, to a ground state acceptor, A. Transitions A and A', B and B', C and C' are coupled and the transitions A, B, C in the donor induce the transitions A', B', C' in the acceptor

pond to the differences in energy between the acceptor vibrational levels connected by the lines labelled A', B' and C' respectively. The transitions represented by the lines A, B, C and A', B', C' are *coupled* transitions and when an excited state donor molecule and a ground state acceptor molecule are a suitable distance apart the decay of the donor to its ground state via the transitions A, B and C will simultaneously induce the promotion of the acceptor to its excited state via the transitions A', B' and C'.

Inspection of the emission spectrum of a potential donor and the absorption spectrum of a potential acceptor will show whether or not there is likely to be a correspondence between the vibrational energy levels of the ground and excited state of the donor and those of the ground and excited state of the acceptor. When the spectra occur over approximately the same wavelength region then there is a high probability of correspondence between the energy levels.

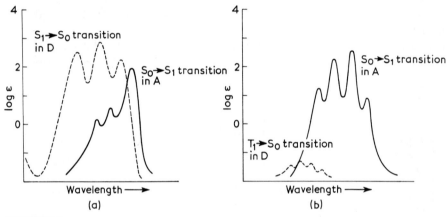

FIGURE 3.7
Overlap of emission spectra of donors, D, with absorption spectra of acceptors, A (a) extensive overlap, both transitions are spin allowed and intense, (b) little overlap, $T_1 \rightarrow S_0$ transition in donor is spin forbidden and weak

The efficiency of energy transfer from the donor to the acceptor is dependent upon the extent of overlap of the emission and absorption spectra of the donor and acceptor respectively and upon whether the decay process $D^* \rightarrow D$ and the excitation process $A \rightarrow A^*$ represent allowed or forbidden electronic transitions. When both are fully allowed transitions and there is extensive overlap of the emission spectrum of D^* and the absorption spectrum of A (as in Fig. 3.7(a)) the energy transfer can be extremely rapid and can exceed the rate at which the donor and acceptor species can diffuse through the medium. If the $D^* \rightarrow D$ process is forbidden (Fig. 3.7(b)) then the transfer will occur at a slower rate, and if the A → A* process is forbidden the rate will be slower still.

Equations have been derived for calculating the rate constant for the energy transfer step and the critical distance for the energy transfer; the critical distance being defined as the intermolecular separation at which the rate of energy transfer equals the rate of spontaneous decay of the excited state. As can be seen from Table 3.5 the calculated values are in reasonable agreement with experimental values.

TABLE 3.5
Critical distances (r) and rate constants (k) for long-range energy transfer in donor-acceptor systems.

Donor	Acceptor	r (nm) Calc.	r (nm) Expt.	k × 10^{-10} (dm^3 mol^{-1} s^{-1}) Calc.	k × 10^{-10} (dm^3 mol^{-1} s^{-1}) Expt.
Anthracene (S_1)	Perylene (S_0)	3.1	5.4	2.3	12
Perylene (S_1)	Rubrene (S_0)	3.8	6.5	2.8	13
Phenanthrene-d$_{10}$(T_1)	Rhodamine B(S_0)	4.5	4.7	–	–
Triphenylamine (T_1)	Fuchsin (S_0)	2.9	3.7	–	–

Energy transfer can occur in principle from the S_1 or T_1 state of the donor to give either an S_1 or T_1 state of the acceptor. However, triplet–triplet energy transfer is unlikely to occur via the long-range mechanism since both transitions involved in this process are spin-forbidden. Also, since an $S_0 \rightarrow S_1$ transition in the acceptor is spin-allowed whereas an $S_0 \rightarrow T_1$ transition is spin-forbidden it is more probable that an S_1 state rather than a T_1 state of the acceptor will be formed on transfer of energy. Thus the most likely long-range transfer processes are singlet–singlet and triplet–singlet energy transfer. These processes are represented by Equations (3.10) and (3.11):

$$D^*(S_1) + A(S_0) \rightarrow D(S_0) + A^*(S_1) \tag{3.10}$$

$$D^*(T_1) + A(S_0) \rightarrow D(S_0) + A^*(S_1) \tag{3.11}$$

Both the transitions $D^*(S_1) \rightarrow D(S_0)$ and $A(S_0) \rightarrow A^*(S_1)$ in the transfer process of Equation (3.10) are spin-allowed and here the energy transfer will occur at large critical distances (5–10 nm) and with a high rate constant ($k = 10^{10}$–10^{11} dm^3 mol^{-1} s^{-1}).

The transition $D^*(T_1) \rightarrow D(S_0)$ in the transfer process of Equation (3.11) is spin-forbidden and hence the energy transfer would generally be expected to occur at a lower rate than the transfer process of equation (3.10). The slower rate for triplet–singlet energy transfer is compensated for by the relatively long lifetime of the T_1 state, and long-range energy transfer from a T_1 state can compete effectively with phosphorescence and $T_1 \leadsto S_0$ intersystem crossing as a means of deactivating the T_1 state of organic molecules.

Kinetics of photochemical processes 4

There are a number of processes, such as intersystem crossing, energy transfer, chemical reaction, whereby absorbed radiant energy can be dissipated by an excited state molecule in its return to the initial ground state. Each of the various processes proceed at different rates for different molecular systems and a knowledge of these rates provides a background to the understanding of the photochemistry of the molecules under study. Each process has an associated rate constant, k, and the evaluation of the rate constants for the different intramolecular and intermolecular processes undergone by photo-excited molecules will be considered in this Chapter.

4.1 Intramolecular processes

Photo-excited molecules are rapidly deactivated to the lowest vibrational level of the first excited singlet state (S_1) and in the absence of chemical reaction and of bimolecular quenching of the excited state the decay to the initial ground state (S_0) will normally occur by intramolecular processes. The possible deactivation processes in the conversion of a molecule from the S_1 state to the S_0 state are shown in Fig. 4.1. The symbols which will be used to identify the rate constant for each process are also given here, as are the orders of magnitude of these constants for organic compounds. It can be seen from Fig. 4.1 that the rate constants for the processes originating from the S_1 state are much larger (10^4–10^{12} s^{-1}) than those for the processes originating from the T_1 state (10^{-2}–10^5 s^{-1}). This reflects the fact that $S_1 \rightarrow S_0$ transitions are spin-allowed whereas $T_1 \rightarrow S_0$ transitions are spin-forbidden.

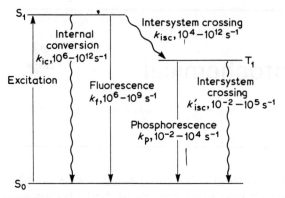

FIGURE 4.1
Intramolecular processes originating from S_1 and T_1 states and the range of magnitude of the corresponding rate constants

4.1.1 Excited State Lifetimes

The lifetime, τ, of an excited state is related to the unimolecular rate constants for the decay processes originating from that state by Equation (3.5). Thus the lifetimes of S_1 and T_1 states deactivated by the intramolecular processes represented in Fig. 4.1 are given by the expressions

$$\tau_{S_1} = \frac{1}{k_{ic} + k_f + k_{isc}} \tag{4.1}$$

$$\tau_{T_1} = \frac{1}{k_p + k'_{isc}} \tag{4.2}$$

Measurement of the lifetimes τ_{S_1} and τ_{T_1} will give information on the rates of the photophysical processes undergone by excited state molecules. When the molecules under study emit fluorescence or phosphorescence the lifetimes can be measured by monitoring the decay of the emitted radiation with time. Phosphorescence lifetimes can be measured using the experimental arrangements shown in Fig. 4.2. In these systems radiation is incident on the sample through either a slit in a rotating cylinder or through a cut-out sector in a rotating chopper. The incident radiation is cut-off from the sample as the cylinder or chopper rotates and there is a pre-set time lag (during which fluorescence is emitted) before the slit in the cylinder or the sector in the second chopper arrives in front of the photomultiplier. The output from the photomultiplier is fed to an oscilloscope or pen recorder and a trace of phosphorescence intensity against time is recorded. The decay of phosphorescence is normally first-order and in such instances the decay will follow the rate law

$$\ln I_0 - \ln I_t = -\frac{t}{\tau} \qquad (4.3)$$

where I_0 and I_t are the phosphorescence intensities at times $t = 0$ and $t = t$ respectively. By measuring the intensity I_t at various times after excitation a straight line plot of $\ln I_t$ or $\log I_t$ against time can be drawn and τ calculated from the reciprocal of the slope of the line. The slope will be the same regardless of the units used for I_t and in practice the photomultiplier output as given on the recorded trace is often used as a measure of I_t.

Fluorescence lifetimes are much shorter than phosphorescence lifetimes and mechanical devices such as those illustrated in Fig. 4.2 cannot be

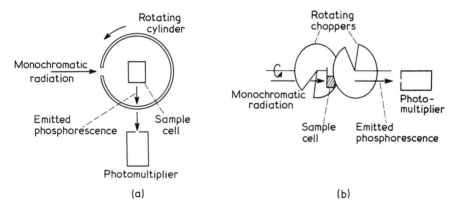

FIGURE 4.2
Apparatus for detecting phosphorescence, (a) plan view of rotating cylinder apparatus, (b) rotating chopper apparatus

operated rapidly enough to enable fluorescence decay to be recorded. A pulse technique is used instead, whereby the S_1 state is populated by excitation with a short duration flash ($\sim 10^{-9}$ s) from a discharge lamp. The exciting flash will be extinguished before the emission of fluorescence is over and the fluorescence decay can be followed using a photomultiplier and oscilloscope arrangement. The fluorescence decay characteristics and the procedure for determining the lifetime of fluorescence are the same as for phosphorescence.

Many photo-excited molecules do not emit fluorescence or phosphorescence and obviously in such cases the lifetimes of the S_1 and T_1 states cannot be determined by methods dependent upon the measurement of emitted radiation. The lifetimes of these states can often be determined, however, using the technique of *flash photolysis*. This technique can be considered with reference to Fig. 4.3.

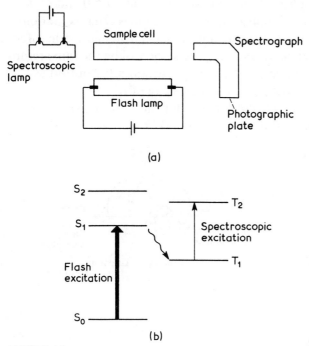

FIGURE 4.3
Flash photolysis, (a) schematic diagram of apparatus, (b) electronic excitation processes occurring in a flash photolysis experiment

In the flash photolysis experiment the sample is subjected to a high intensity flash of short duration ($\sim 10^{-9}$–10^{-6} s) which excites a high proportion of the ground state molecules into the S_1 state. If the quantum yield for $S_1 \rightsquigarrow T_1$ intersystem crossing is non-zero then within a period of $\sim 10^{-12}$–10^{-4} s the T_1 state of the molecule will be populated. The T_1 triplet state is relatively long-lived ($\sim 10^{-5} - 10^2$ s) and during its lifetime a low intensity flash from a second lamp is triggered off. The second lamp is referred to as the spectroscopic lamp since the radiation emitted from this lamp passes through the sample cell and into a spectrograph where the spectrum of the radiation is recorded on a photographic plate. If the radiation from the spectroscopic lamp covers the wavelength range for the triplet–triplet transitions ($T_1 \rightarrow T_2$, $T_1 \rightarrow T_3$ etc.) and if the ground state molecules do not absorb in this range then a triplet → triplet absorption band or bands will be observed in the spectrum of the lamp radiation. (Triplet → triplet transitions are spin-allowed transitions and the corresponding absorption bands can be of high intensity.) Once the existence and wavelength position of the triplet → triplet absorption has been ascertained the lifetime of the T_1 state can be determined by measuring the decay in intensity of the absorption with time. In order to do this the spectroscopic lamp is replaced by a conventional xenon or tungsten lamp, and the spectrograph is replaced by a monochromator and photomultiplier-oscilloscope assembly. The monochromator is normally set to the wavelength of maximum absorption in the $T_1 \rightarrow T_2$ band. On flash excitation there will be an immediate decrease in the intensity of the radiation from the xenon or tungsten lamp reaching the photomultiplier because of the onset of the $T_1 \rightarrow T_2$ absorption. As molecules in the T_1 state decay to the S_0 state so the $T_1 \rightarrow T_2$ absorption will decrease until the original level of intensity is regained. The output from the photomultiplier is fed to the oscilloscope and the change in output with time as shown on the screen of the oscilloscope is recorded photographically. The trace on the photograph will show the decay of the T_1 state with time. For molecules which phosphoresce the decay curve obtained by flash photolysis will be equivalent to the decay curve obtained by the measurement of phosphorescence emission if both are recorded under the same experimental conditions. Decay curves obtained by flash photolysis are analyzed in the same manner as phosphorescence decay curves to give the lifetime of the T_1 state of the molecule.

The lifetimes of S_1 states can also, in principle, be obtained using the flash photolysis technique. However S_1 states are very short-lived ($\sim 10^{-9} - 10^{-6}$ s) and extremely short duration excitation flashes ($\sim 10^{-9}$ s) and rapid detection systems are required to observe transitions originating from S_1 states. Lasers have been found to be suitable as flash sources and as trigger devices in the monitoring system and flash photolysis has now been used to measure the lifetime of the S_1 state of a number of organic compounds. Some of the

TABLE 4.1
Lifetimes of singlet states (*nanoseconds*) and of triplet states (*seconds*) as measured by emission methods and by flash photolysis (F.P.)

Compound	S_1 state[a]		T_1 state[b]	
	Emission	F.P.	Emission	F.P.
Pyrene	261	296	0.2	0.7
Phenanthrene	67.2[c]	65.0[c]	3.3	3.3
1,2-benzanthracene	44.0	45.0	–	–
3,4-benzpyrene	57.5	49.1	–	–
Naphthalene	–	–	2.3	3.3
Anthracene	–	–	0.1	0.1

[a] Measurements made in cyclohexane solution.
[b] Measurements made in ether – isopentane – ethanol at 77° K.
[c] Measurements made in rigid polymethylmethacrylate.

results are given in Table 4.1 along with the corresponding values obtained by measurement of the fluorescence emission. Triplet state lifetimes as determined by flash photolysis and by the phosphorescence emission method are also included in Table 4.1. It can be seen that the results obtained by the two methods — flash photolysis and emission — are in agreement for both singlet and triplet states.

The *natural* radiative lifetime of an excited state is the emission lifetime of that state when there are no radiationless processes contributing to its decay. The *observed* radiative lifetime, which is the experimentally measured lifetime of fluorescence or phosphorescence, will usually be shorter than the natural radiative lifetime because of the occurrence of radiationless decay processes. When this is so, the natural radiative lifetimes, τ_f^o and τ_p^o, may be determined from the relationships

$$\tau_f^o = \frac{\tau_f}{\phi_f} \quad : \quad \tau_p^o = \frac{\tau_p}{\phi_p} \tag{4.4}$$

where τ_f and τ_p are the experimental fluorescence and phosphorescence lifetimes respectively, and ϕ_f and ϕ_p are the respective quantum yields of fluorescence and phosphorescence. The quantum yields of fluorescence and phosphorescence can be readily measured using the appropriate apparatus. However a description of the instrumentation required and the methods of calculation is beyond the scope of the present text (see Bibliography).

The natural radiative lifetimes of S_1 and T_1 states can sometimes be determined from absorption spectra. For example, if the natural radiative lifetime of the S_1 state is required then the mean wave number, $\bar{\nu}$, of the $S_0 \rightarrow S_1$ absorption band and the integrated absorption intensity, $\int \epsilon \, d\bar{\nu}$, of this band are measured. These values are then substituted into the following equation† to obtain an approximate value for the radiative lifetime:

$$\tau^\circ = \frac{3.47 \times 10^8}{\bar{\nu}^2 \int \epsilon \, d\bar{\nu}} \cdot \frac{g_u}{g_l} \tag{4.5}$$

where g_u and g_l are respectively the degeneracies of the upper and lower electronic states involved in the transition. The degeneracy terms are equal to one for a singlet state and three for a triplet state.

An approximate value for the natural radiative lifetime of a T_1 state can also be found from Equation (4.5), but this time the value for $\bar{\nu}$ and $\int \epsilon \, d\bar{\nu}$ to be substituted into the equation are obtained from the $S_0 \rightarrow T_1$ absorption band. Since the $S_0 \rightarrow T_1$ transition is spin-forbidden the corresponding $S_0 \rightarrow T_1$ band will be weak and there is likely to be a large uncertainty in the measured value of the integrated absorption intensity. As can be seen from the results given in Table 4.2 the lifetimes of T_1 states determined by this method do not agree so closely with the lifetimes determined by the emission method as do the lifetimes of S_1 states.

4.1.2 *Fluorescence and phosphorescence*

The emission of fluorescence or phosphorescence by an excited state molecule is normally a unimolecular first order process and the equations describing the rate of the emission take the following form:

$$\text{rate of fluorescence} = k_f [S_1] \tag{4.6}$$

$$\text{rate of phosphorescence} = k_p [T_1] \tag{4.7}$$

†See *J. G. Calvert* and *J. N. Pitts, Jr.*, "Photochemistry" (Wiley, New York, 1966) pp. 170–174.

INTRODUCTION TO MOLECULAR PHOTOCHEMISTRY

TABLE 4.2
Comparison of natural radiative lifetimes of singlet states (*nanoseconds*) and of triplet states (*milliseconds*) as determined from absorption spectra and from radiative emission.

Compound	S_1 state		T_1 state	
	Absorption[a]	Emission[b]	Absorption[c]	Emission[b]
Anthracene	13.5	16.7	90.0	0.1
Perylene	5.1	5.6	–	–
Rubrene	22.0	16.0	–	–
Bromobenzene	–	–	3.0	1.0
2-bromonaphthalene	–	–	10.0	20.0

[a] Derived from measurements on the $S_0 \rightarrow S_1$ absorption band.
[b] Calculated using Equation (4.4).
[c] Derived from measurements on the $S_0 \rightarrow T_1$ absorption band.

where k_f and k_p are the rate constants for the fluorescence and phosphorescence processes respectively, and $[S_1]$ and $[T_1]$ are the concentrations of the excited state molecules giving rise to the radiative emission.

The rate constants for the fluorescence and phosphorescence processes can be derived from the experimental values of τ_f, ϕ_f and τ_p, ϕ_p respectively via the following equations:

$$k_f = \frac{1}{\tau_f^0} = \frac{\phi_f}{\tau_f} \tag{4.8}$$

$$k_p = \frac{1}{\tau_p^0} = \frac{\phi_p}{\tau_p} \tag{4.9}$$

Values of the rate constants for fluorescence and phosphorescence emission in some organic compounds are given in Table 4.3. It can be seen that for a given molecule the rate constant for the fluorescence process is several orders of magnitude greater than that for the phosphorescence process.

4.1.3 Intersystem crossing

The rate constants for the $S_1 \rightsquigarrow T_1$ and $T_1 \rightsquigarrow S_0$ intersystem crossings can sometimes be calculated from the experimental values of the rate constants and quantum yields of fluorescence and phosphorescence emission. The

KINETICS OF PHOTOCHEMICAL PROCESSES

Table 4.3
Rate constants for fluorescence emission and phosphorescence emission.

Compound	k_f (s^{-1})	k_p (s^{-1})
Naphthalene	1×10^6	1.6×10^{-2}
1-chloronaphthalene	3×10^6	5.7×10^{-1}
Benzene	2×10^6	3.5×10^{-2}
Benzophenone	1×10^6	1.6×10^2

equations relating the rate constants for the intersystem crossings to the rate constants and quantum yields of fluorescence and phosphorescence can be derived by a kinetic analysis of the reaction scheme representing excitation and decay of the molecule. The scheme for the excitation and decay processes shown in Fig. 4.1 may be written as follows:

Process	Representation	Rate
Excitation	$S_0 + h\nu \to S_1$	I
Internal conversion	$S_1 \to S_0$	$k_{ic}[S_1]$
Fluorescence	$S_1 \to S_0 + h\nu_f$	$k_f[S_1]$
Intersystem crossing	$S_1 \to T_1$	$k_{isc}[S_1]$
Intersystem crossing	$T_1 \to S_0$	$k'_{isc}[T_1]$
Phosphorescence	$T_1 \to S_0 + h\nu_p$	$k_p[T_1]$

where I is the rate of absorption of the incident radiation.

According to the *stationary state* or *steady state* hypothesis the concentrations of the intermediates in the reaction will reach a stationary or steady value soon after irradiation commences and from then on, while irradiation continues, the rate of formation of the intermediates will equal their rate of removal, provided the concentration of the absorbing solute is such that all the incident radiation is absorbed during the period of irradiation. The

INTRODUCTION TO MOLECULAR PHOTOCHEMISTRY

intermediates in the reaction sequence shown above are molecules in the S_1 and T_1 states and the mathematical expressions of the hypothesis relating to these intermediates are:

(a) rate of formation of S_1 = rate of removal of S_1

$$I = k_{ic}[S_1] + k_f[S_1] + k_{isc}[S_1] \tag{4.10}$$

(b) rate of formation of T_1 = rate of removal of T_1

$$k_{isc}[S_1] = k'_{isc}[T_1] + k_p[T_1] \tag{4.11}$$

Equation (4.10) and (4.11) can be written in alternative form:

$$[S_1] = \frac{I}{k_{ic} + k_f + k_{isc}} \tag{4.12}$$

$$[T_1] = \frac{k_{isc}[S_1]}{k'_{isc} + k_p} \tag{4.13}$$

Substitution for $[S_1]$ into Equation (4.13) gives

$$[T_1] = \frac{k_{isc} I}{(k'_{isc} + k_p)(k_{ic} + k_f + k_{isc})} \tag{4.14}$$

Now the quantum yields for fluorescence and for phosphorescence are given by the equations:

$$\phi_f = \frac{\text{rate of fluorescence emission}}{\text{rate of absorption of radiation}} = \frac{k_f[S_1]}{I} \tag{4.15}$$

$$\phi_p = \frac{\text{rate of phosphorescence emission}}{\text{rate of absorption of radiation}} = \frac{k_p[T_1]}{I} \tag{4.16}$$

Therefore:

$$\frac{\phi_p}{\phi_f} = \frac{k_p[T_1]}{k_f[S_1]} \tag{4.17}$$

Substituting for $[S_1]$ and $[T_1]$ in Equation (4.17) gives:

$$\frac{\phi_p}{\phi_f} = \frac{k_p}{k_f}\left(\frac{k_{isc}}{k'_{isc} + k_p}\right) \tag{4.18}$$

KINETICS OF PHOTOCHEMICAL PROCESSES

If the total emission yield is high, i.e. $\phi_f + \phi_p \to 1$, then decay by $T_1 \rightsquigarrow S_0$ intersystem crossing will be minimal. Consequently k_p will be much greater than k'_{isc} and Equation (4.18) will reduce to:

$$\frac{\phi_p}{\phi_f} = \frac{k_{isc}}{k_f} \qquad (4.19)$$

Thus measurement of the quantities ϕ_f, ϕ_p and k_f for the molecule under study will allow a value to be obtained for the rate constant, k_{isc}, for the $S_1 \rightsquigarrow T_1$ intersystem crossing process.

The rate constant, k'_{isc}, for the $T_1 \rightsquigarrow S_0$ intersystem crossing, can be calculated from the values of ϕ_f, ϕ_p and k_f for systems where the excited state molecules, which do not emit radiation, decay solely via the T_1 state. The results given in Table 3.1 show that the sum $\phi_f + \phi_{isc}(S_1 \rightsquigarrow T_1)$ for the molecules listed therein is close to unity, i.e. the decay of the excited states of these molecules occurs either by fluorescence or by processes originating from the T_1 state. This situation is represented mathematically by Equation (4.20):

$$\phi_f + \phi_p + \phi_{isc}(T_1 \rightsquigarrow S_0) = 1 \qquad (4.20)$$

Now the quantum yield for the $T_1 \rightsquigarrow S_0$ intersystem crossing process is given by:

$$\phi_{isc}(T_1 \rightsquigarrow S_0) = \frac{\text{rate of } T_1 \rightsquigarrow S_0 \text{ intersystem crossing}}{\text{rate of absorption of radiation}} = \frac{k'_{isc}[T_1]}{I} \qquad (4.21)$$

Combining Equations (4.16) and (4.21) gives:

$$\frac{\phi_p}{\phi_{isc}(T_1 \rightsquigarrow S_0)} = \frac{k_p}{k'_{isc}} \qquad (4.22)$$

Rearranging Equation (4.22) gives:

$$\phi_{isc}(T_1 \rightsquigarrow S_0) = \frac{k'_{isc}}{k_p} \cdot \phi_p \qquad (4.23)$$

Substituting for $\phi_{isc}(T_1 \rightsquigarrow S_0)$ from Equation (4.23) into Equation (4.20) results in the equation:

$$k'_{isc} = k_p \frac{1 - (\phi_f + \phi_p)}{\phi_p} \qquad (4.24)$$

Thus k'_{isc} can be calculated from the values of k_p, ϕ_p and ϕ_f. Once a value is known for k'_{isc} the value of k_{isc} can be obtained from Equation (4.18) for those molecules which do not have a high emission yield.

TABLE 4.4
Rate constants for $S_1 \leadsto T_1$ and $T_1 \leadsto S_0$ intersystem crossing processes.

Compound	$k_{isc}(S_1 \leadsto T_1)^a$ (s^{-1})	$k'_{isc}(T_1 \leadsto S_0)^b$ (s^{-1})
Naphthalene	8×10^5	0.12
1-chloronaphthalene	5×10^7	0.42
Benzene	8×10^6	0.10
Benzophenone	1×10^{10}	18.0
Acetophenone	5×10^9	50.0

[a] Calculated from Equation (4.19).
[b] Calculated from Equation (4.24).

Values for the rate constants of the intersystem crossing in a number of compounds are listed in Table 4.4. A knowledge of the rate constant for the $S_1 \rightarrow T_1$ intersystem crossing is of importance since the value indicates whether a molecule is liable to react photochemically from the S_1 state or from the T_1 state. The rate constant, k_{isc}, for the $S_1 \rightarrow T_1$ intersystem crossing in aromatic ketones is very high ($\sim 10^9 - 10^{10}$ s^{-1}) and consequently the photoreactive state of aromatic ketones is normally the T_1 state. On the other hand, the rate constants for the $S_1 \rightarrow T_1$ intersystem crossing in aromatic hydrocarbons are much lower ($\sim 10^6 - 10^7$ s^{-1}) than in aromatic ketones and therefore there is a much greater likelihood of aromatic hydrocarbons reacting while in the S_1 state.

4.1.4 Internal conversion

The rate constant for the $S_1 \rightarrow S_0$ internal conversion in an excited state molecule can be calculated from Equation (4.1) provided the fluorescence lifetime of the S_1 state and the rate constants k_f and k_{isc} are known. Rate constants for $S_1 \rightarrow S_0$ internal conversion in organic molecules normally lie in the range $10^6 - 10^{12}$ s^{-1}.

4.2 Intermolecular processes

The deactivation of excited state molecules via intermolecular interactions is a frequent occurrence in photochemical systems. The major routes for the dissipation of the excitation energy in the photo-excited species are (a) energy transfer to an acceptor molecule, and (b) chemical reaction with a suitable reactant. Intermolecular and intramolecular deactivation processes occur concurrently and all deactivation steps have to be taken into account when considering the kinetics of a photochemical system. The processes which can contribute to the deactivation of an excited state molecule and which may have to be considered in a kinetic analysis are given in Fig. 4.4. The

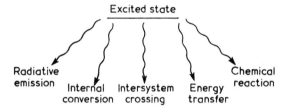

FIGURE 4.4
Possible paths for the deactivation of an excited state

rate constants for the intramolecular processes — radiative emission, internal conversion, intersystem crossing — are normally evaluated under experimental conditions such that energy transfer and chemical reaction are eliminated. It is obviously not possible to select experimental conditions such that intramolecular deactivation processes are eliminated and rate terms for these processes may come into the equations used to derive the rate constants for energy transfer and chemical reaction.

4.2.1 *Energy transfer*

Energy transfer from a photo-excited donor, D^*, to the ground state of an acceptor, A, normally takes place from either the S_1 or the T_1 state of the donor with the formation of an S_1 or a T_1 state of the acceptor. That

INTRODUCTION TO MOLECULAR PHOTOCHEMISTRY

is, either singlet–singlet, triplet–singlet or triplet–triplet energy transfer may occur. Only singlet–singlet and triplet–triplet energy transfer will be considered in this section. Energy transfer is often referred to as *quenching* since if it is sufficiently rapid all excited state molecules will be deactivated by this route and the fluorescence or phosphorescence emitted from the excited states will be quenched.

(a) Triplet–triplet energy transfer. Triplet–triplet energy transfer can occur when the relative energies of the lowest excited states of the donor and acceptor are as in Fig. 4.5. Transfer from the S_1 state of the donor to the S_1 and T_1 states of the acceptor are respectively energy- and spin-forbidden and are inefficient processes as compared to the $T_1 \rightarrow T_1$ transfer which is both energy- and spin-allowed for the situation shown in Fig. 4.5. Similarly, transfer from the T_1 state of the donor to the S_1 state of the acceptor will be highly inefficient since this process is both energy- and spin-forbidden.

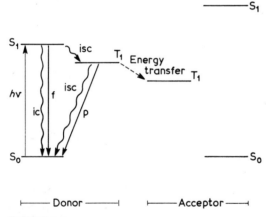

FIGURE 4.5
Favourable arrangement of donor–acceptor energy levels for triplet–triplet energy transfer

The various intramolecular processes which compete with the triplet–triplet energy transfer in deactivating the photo-excited donor are shown in Fig. 4.5. (It is assumed here that the photo-excited donor does not take part in a chemical reaction). The reaction scheme representing the processes given in Fig. 4.5 can be written as follows:

KINETICS OF PHOTOCHEMICAL PROCESSES

Process	Representation	Rate
Excitation	$S_0(D) + h\nu \to S_1(D)$	I
Internal conversion	$S_1(D) \to S_0(D)$	$k_{ic}[S_1]$
Fluorescence	$S_1(D) \to S_0(D) + h\nu_f$	$k_f[S_1]$
Intersystem crossing	$S_1(D) \to T_1(D)$	$k_{isc}[S_1]$
Phosphorescence	$T_1(D) \to S_0(D) + h\nu_p$	$k_p[T_1]$
Intersystem crossing	$T_1(D) \to S_0(D)$	$k'_{isc}[T_1]$
Energy transfer (quenching)	$T_1(D) + S_0(A) \to S_0(D) + T_1(A)$	$k_e[T_1][A]$

where k_e is the rate constant for triplet–triplet energy transfer.

Applying the stationary state hypothesis to the intermediate states S_1 and T_1 of the donor gives Equations (4.25) and (4.26) respectively:

$$I = k_{ic}[S_1] + k_f[S_1] + k_{isc}[S_1] \qquad (4.25)$$

$$k_{isc}[S_1] = k_p[T_1] + k'_{isc}[T_1] + k_e[T_1][A] \qquad (4.26)$$

Rearrangement of Equations (4.25) and (4.26) gives:

$$[S_1] = \frac{I}{k_{ic} + k_f + k_{isc}} \qquad (4.27)$$

$$[T_1] = \frac{k_{isc}[S_1]}{k_p + k'_{isc} + k_e[A]} \qquad (4.28)$$

Now the quantum yields of phosphorescence in the presence and absence of the energy acceptor, A, are given by Equations (4.29) and (4.30) respectively:

$$\phi_p = \frac{k_p[T_1]}{I} \qquad (4.29)$$

$$\phi_p^0 = \frac{k_p[T_1]'}{I} \qquad (4.30)$$

where $[T_1]'$ is the stationary state concentration of triplet state donor in the absence of the energy acceptor A. This concentration is given by Equation (4.14). Combining Equations (4.29) and (4.30) gives:

$$\frac{\phi_p^0}{\phi_p} = \frac{[T_1]'}{[T_1]} \tag{4.31}$$

Substituting for $[T_1]'$ from Equation (4.14) and $[T_1]$ from Equation (4.28) into Equation (4.31) gives:

$$\frac{\phi_p^0}{\phi_p} = \frac{k_p + k'_{isc} + k_e[A]}{k_p + k'_{isc}} \tag{4.32}$$

Rearrangement of Equation (4.32) gives:

$$\frac{\phi_p^0}{\phi_p} = 1 + \frac{k_e}{k_p + k'_{isc}} \cdot [A] \tag{4.33}$$

If the ratio of the intensities of the phosphorescence emission in the absence and presence of the acceptor is measured instead of the ratio ϕ_p^0/ϕ_p then Equation (4.34) below is used in place of Equation (4.33):

$$\frac{I_p^0}{I_p} = 1 + \frac{k_e}{k_p + k'_{isc}} \cdot [A] \tag{4.34}$$

Equation (4.34) shows that a plot of the phosphorescence intensity ratio I_p^0/I_p against [A] will be a straight line of slope equal to $\frac{k_e}{k_p + k'_{isc}}$. Thus if k_p and k'_{isc} are known for the donor, the value of the rate constant, k_e, for triplet–triplet energy transfer can be calculated from the slope of the line.

Equations of the form of (4.33) and (4.34) are known as *Stern–Volmer* equations and the plots as *Stern–Volmer* plots.

If the T_1 state of the donor is deactivated solely by the intramolecular processes of phosphorescence and $T_1 \rightsquigarrow S_0$ intersystem crossing then the lifetime of the T_1^* state is given by:

$$\tau = \frac{1}{k_p + k'_{isc}} \tag{4.35}$$

When the T_1 state is additionally deactivated by triplet–triplet energy transfer the expression for the lifetime, τ^e, will take the form:

$$\tau^e = \frac{1}{k_p + k'_{isc} + k_e[A]} \tag{4.36}$$

Combining Equations (4.35) and (4.36) results in Equation (4.37)

$$\frac{1}{\tau^e} = \frac{1}{\tau} + k_e[A] \tag{4.37}$$

Equation (4.37) shows that a plot of the reciprocal of the lifetime, τ^e, of the T_1 state against the concentration, [A], of the acceptor should be a straight line of slope k_e. Equation (4.37) holds for the triplet state of biacetyl in fluid solution, and some values derived from this equation for the rate constant of energy transfer from the biacetyl triplet state to different acceptors are given in Table 4.5.

TABLE 4.5
Rate constants for triplet–triplet energy transfer *to* and *from* the T_1 state of biacetyl (E_T = 234 kJ mol^{-1}) in benzene solution at 20°.

	E_T (kJ mol^{-1})	Rate constants for energy transfer to biacetyl (dm^3 mol^{-1} s^{-1})	Rate constants for energy transfer from biacetyl (dm^3 mol^{-1} s^{-1})
Naphthalene	255	1 × 10^{10}	2 × 10^6
1-chloronaphthalene	246	4 × 10^9	3 × 10^7
2,2'-dinaphthyl	234	1 × 10^9	3 × 10^9
Fluoranthene	226	2 × 10^7	5 × 10^9
1,2-benzpyrene	226	5 × 10^7	6 × 10^9
Pyrene	205	2 × 10^4	8 × 10^9

Biacetyl can act as an energy acceptor as well as an energy donor and if another molecular species is selectively excited in the presence of biacetyl then transfer of energy to ground state biacetyl may occur. The rate constants for the transfer of triplet state energy to biacetyl have been evaluated in a number of systems and some of these values are given in Table 4.5.

It can be seen from Table 4.5 that the values for the rate constants of triplet–triplet energy transfer approach 10^{10} dm^3 mol^{-1} s^{-1} when the energy of the lowest triplet state of the acceptor is below that of the triplet state of the donor. As the energy of the triplet state of the acceptor increases relative to that of the donor so the value for the rate constant drops. The limiting value of 10^{10} dm^3 mol^{-1} s^{-1} is approximately the same as that calculated from the Debye equation (page 50) for the rate constant of a diffusion-controlled bimolecular reaction. This indicates that energy transfer to acceptors with low energy T_1 states takes place at every collision of donor and acceptor. If the acceptor molecule, to which energy is transferred, collides with a ground state donor molecule before it is deactivated then the energy may be transferred back to the donor, i.e. reversible energy transfer can occur. Table 4.5 shows, for example, that for the biacetyl-fluoranthene system the rate constant for biacetyl to fluoranthene triplet energy transfer is 5×10^9 dm^3 mol^{-1} s^{-1} while that for the reverse transfer is 2×10^7 dm^3 mol^{-1} s^{-1}.

The procedures described above for determining the rate constants for triplet–triplet energy transfer between donors and acceptors depend upon the observation and measurement of emitted phosphorescence. However, since the majority of molecules do not phosphoresce in fluid solution the technique of flash photolysis frequently has to be used to gain information on triplet energy transfer. Evidence for triplet energy transfer can often be obtained by selectively flash-exciting a donor in the presence of an acceptor and comparing the absorption spectrum of the mixture, at suitable times after flashing, with that of a flash-excited solution of donor alone. For example, when phenanthrene is flash-excited in the presence of naphthalene the absorption spectrum of the solution immediately after flashing shows only the naphthalene triplet–triplet absorption, whereas if a solution of phenanthrene alone is flash-excited the absorption spectrum shows only the phenanthrene triplet–triplet absorption. These observations show unequivocally that the triplet state of phenanthrene is rapidly quenched by naphthalene and that triplet state naphthalene is formed as a result of the quenching. That is to say, energy is transferred from the triplet state phenanthrene to ground state naphthalene and the transferred energy is used to form triplet state naphthalene.

Rate constants for triplet–triplet energy transfer can be obtained from flash photolytic measurements of the rate of decay of the T_1 state of the

KINETICS OF PHOTOCHEMICAL PROCESSES

donor molecule. At low concentrations of donor, when there is no self-quenching of triplet states, the T_1 state will be deactivated by phosphorescence and intersystem crossing (intramolecular process) and by triplet–triplet energy transfer to the acceptor (intermolecular process). Thus the rate of decay of the T_1 state will equal the sum of the rates of the intramolecular and intermolecular deactivation processes. This is expressed by the equation:

$$-\frac{d[T_1]}{dt} = k_d [T_1] + k_e [T_1][A] \tag{4.38}$$

where k_d is the overall rate constant for intramolecular deactivation of the T_1 state. The rate constant k_d is equal to the reciprocal of the lifetime of the T_1 state in the absence of acceptor and is obtained from measurements of this lifetime using the flash photolysis technique.

An indirect method of measuring the rate constant k_e is to determine the concentration, $[A]$, of added acceptor necessary to double the rate of decay of the T_1 state relative to when there is no acceptor present. For this particular concentration the terms on the right-hand side of Equation (4.38) will have the same value, i.e. Equation (4.39) will be applicable:

$$k_d [T_1] = k_e [T_1][A] \tag{4.39}$$

That is:

$$k_e = \frac{k_d}{[A]} \tag{4.40}$$

Thus if k_d is measured in the absence of acceptor, then k_e can be found from Equation (4.40). Values of k_e determined by this method are given in Table 4.6.

A direct method for evaluating the rate constant k_e is to make measurements of the *absolute* rate of decay of the triplet state in both the presence and absence of an energy acceptor. If the concentration $[A]$ of acceptor is arranged to be high, the rate of energy transfer, $k_e [T_1][A]$, will be pseudo-first order and will be given by the expression

$$\text{rate} = k[T_1] \tag{4.41}$$

where k is the rate constant for energy transfer in the presence of acceptor. This rate constant is equal to the reciprocal of the observed lifetime of the

TABLE 4.6.
Rate constants for triplet–triplet energy transfer as measured by flash photolysis.

Donor	Acceptor	k_e (dm³ mol⁻¹ s⁻¹)	Energy gap $D(T_1) - A(T_1)$ (kJ mol⁻¹)
Phenanthrene	Naphthalene	2.9×10^6	5
Naphthalene	1-iodonaphthalene	2.8×10^8	9
Triphenylene	Naphthalene	1.3×10^9	24
Benzophenone	Naphthalene	1.2×10^9	36

T_1 state in the presence of acceptor. Since the lifetime will be extremely short when a high concentration of A is present the laser flash technique has to be used for its measurement. Once the values of k and k_d have been determined the value of k_e can be calculated from the equation:

$$k = k_d + k_e[A] \tag{4.42}$$

Values obtained from measurements of the decay of the benzophenone triplet state in the presence and absence of naphthalene are given in the following equation:

$$4.3 \times 10^6 \text{ s}^{-1} = 4.0 \times 10^5 \text{ s}^{-1} + k_e[8.15 \times 10^{-4} \text{ mol l}^{-1}] \tag{4.43}$$

Solving Equation (4.43) gives a value of 4.8×10^9 dm³ mol⁻¹ s⁻¹ for the rate constant, k_e, for triplet–triplet energy transfer from benzophenone to naphthalene. This compares reasonably well with the value obtained by the indirect method (see Table 4.6).

The results given in Table 4.6 show that the rate constants for energy transfer increase as the energy of the acceptor triplet state falls below that of the donor triplet state and that the limiting value approaches that of the diffusion-controlled rate constant. This is in agreement with results obtained from measurement of phosphorescence emission (Table 4.5).

The phenomenon of triplet–triplet energy transfer can sometimes be used as a means of determining the rate of the $S_1 \rightsquigarrow T_1$ intersystem crossing in the donor. The principle of the method is as follows. An acceptor is chosen which undergoes a photo-reaction via its T_1 state (e.g. the *cis – trans* isomerization of olefins) and the donor is used to sensitize this reaction by transfer of

KINETICS OF PHOTOCHEMICAL PROCESSES

triplet state energy. It is arranged that the T_1 level of the acceptor is sufficiently below the T_1 level of the donor so that the transfer will be diffusion-controlled. Under this circumstance, the rate of population of the T_1 state of the acceptor will be equal to the rate at which the T_1 state of the donor is populated, i.e. equal to the rate of the intersystem crossing in the donor. Thus, by measuring the rate of the chemical process undergone by the acceptor, the rate (or quantum yield) of the $S_1 \rightsquigarrow T_1$ intersystem crossing in the donor can be found. Values of ϕ_{isc} obtained by this method are given on page 44.

(b) Singlet–singlet energy transfer. Rate constants for singlet–singlet energy transfer can be evaluated in certain instances from measurements made on the fluorescence emitted by the donor or acceptor species. Biacetyl has been used frequently in such studies because (a) biacetyl fluoresces in solution, and (b) the S_1 state of biacetyl is of relatively low energy and hence biacetyl can act as an efficient quencher of donor-excited singlet states. The procedures used in determining rate constants for singlet–singlet energy transfer can be considered in relation to the energy transfer between naphthalene as the donor, D, and biacetyl as the acceptor, A. The relative energies of the lowest excited states of naphthalene and biacetyl are shown in Fig. 4.6 along

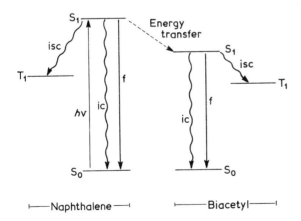

FIGURE 4.6
Decay processes originating from the S_1 states of naphthalene and biacetyl after selective excitation of naphthalene

INTRODUCTION TO MOLECULAR PHOTOCHEMISTRY

with the deactivation processes originating from the naphthalene and biacetyl S_1 states. The deactivation steps originating from the T_1 states are not shown since these do not affect the kinetics of the singlet state processes. In practice, the experimental solutions are saturated with oxygen so that the T_1 states are rapidly deactivated and the phosphorescence emission of biacetyl completely quenched.

The reaction scheme corresponding to the processes shown in Fig. 4.6 is as follows:

Process	Representation	Rate
Excitation	$S_0(D) + h\nu \rightarrow S_1(D)$	I
Internal conversion	$S_1(D) \rightarrow S_0(D)$	$k_{ic}[S_1]_D$
Fluorescence	$S_1(D) \rightarrow S_0(D) + h\nu_f$	$k_f[S_1]_D$
Intersystem crossing	$S_1(D) \rightarrow T_1(D)$	$k_{isc}[S_1]_D$
Energy transfer	$S_1(D) + S_0(A) \rightarrow S_0(D) + S_1(A)$	$k_e[S_1]_D[A]$
Internal conversion	$S_1(A) \rightarrow S_0(A)$	$k'_{ic}[S_1]_A$
Fluorescence	$S_1(A) \rightarrow S_0(A) + h\nu'_f$	$k'_f[S_1]_A$
Intersystem crossing	$S_1(A) \rightarrow T_1(A)$	$k'_{isc}[S_1]_A$

Applying the stationary state treatment to the S_1 state of the donor leads to the following expressions:

$$[S_1]_D = \frac{I}{k_{ic} + k_f + k_{isc} + k_e[A]} \tag{4.44}$$

$$\phi_{f(D)} = \frac{k_f}{k_{ic} + k_f + k_{isc} + k_e[A]} \tag{4.45}$$

When there is no acceptor present Equation (4.45) becomes:

$$\phi^0_{f(D)} = \frac{k_f}{k_{ic} + k_f + k_{isc}} \tag{4.46}$$

Thus the ratio of the fluorescence quantum yields in the absence and in the presence of acceptor is given by:

$$\frac{\phi^0_{f(D)}}{\phi_{f(D)}} = 1 + \frac{k_e}{k_{ic} + k_f + k_{isc}} \cdot [A] \qquad (4.47)$$

When the values of the rate constants k_{ic}, k_f and k_{isc} are known for the donor the rate constant, k_e, for energy transfer from the donor singlet state to the acceptor can be calculated from the slope of the plot of $\phi^0_{f(D)}/\phi_{f(D)}$ against [A]. Values of the rate constants for energy transfer from the singlet state of

TABLE 4.7
Rate constants for singlet–singlet energy transfer from aromatic hydrocarbons to biacetyl.

Compound	Energy of S_1 state (kJ mol^{-1})	k_e (dm^3 mol^{-1} s^{-1})
Biacetyl	277	–
Benzene[a]	458	3.3×10^{10}
Toluene[a]	453	3.7×10^{10}
Biphenyl[b]	400	2.2×10^{10}
Naphthalene[a]	380	2.2×10^{10}

[a] Aerated hexane solutions at 28°C.
[b] Aerated hexane solutions at 25°C.

a number of donors to biacetyl are given in Table 4.7. These values are similar to the diffusion-controlled rate constant which for hexane at 28°C is calculated to be 2.4×10^{10} dm^3 mol^{-1} s^{-1}. The high values for the measured rate constants arise because of the large energy gap between the S_1 state of the donors and the S_1 state of biacetyl.

The system represented in Fig. 4.6 may also be considered from the viewpoint of sensitization of the fluorescence of biacetyl. If the fluorescence spectrum of the acceptor lies in a different region of the spectrum from that of the donor then the intensity, $I_{f(A)}$, of the acceptor fluorescence may be measured. Also, if the acceptor does not absorb any of the incident radiation the reaction scheme can be analyzed to yield the following expression:

$$\frac{I_{f(A)}}{I_{f(D)}} = \frac{k'_f}{k_f} \cdot \frac{k_e}{k'_{ic} + k'_f + k'_{isc}} [A] \qquad (4.48)$$

Thus by measuring the ratio $I_{f(A)}/I_{f(D)}$ when known concentrations of the energy acceptor are present a value of k_e can be derived, provided the various rate constants for the intramolecular processes represented in Equation (4.48) are known. The values obtained for k_e from Equation (4.48) can be used to check the value obtained using Equation (4.47).

The experimental values for the rate constants of singlet–singlet and triplet–triplet energy transfer are of importance in ascertaining whether the transfer occurs by the short-range or the long-range mechanism (pages 51 and 55). The observation that the rate constants for the singlet–singlet and triplet–triplet energy transfer processes discussed earlier approach, but do not exceed, the diffusion-controlled rate constant indicates that the mechanism involves collision or near collision of the donor and acceptor species. That is, the energy transfer occurs by the short-range mechanism. On the other hand, the values for the rate constants for singlet–singlet transfer between the donors and acceptors listed in Table 4.8 are about an order of magnitude greater than the calculated diffusion-controlled rate constant ($\sim 10^{10}$ dm^3 mol^{-1} s^{-1}). This indicates that the energy transfer in these particular systems must be by the long-range mechanism. Also, the values for the rate constants quoted in Table 4.8 do not alter when the viscosity of the

TABLE 4.8
Rate constants for singlet–singlet energy transfer in some donor–acceptor systems.

Donor	Acceptor	k_e (dm^3 mol^{-1} s^{-1})
1-chloroanthracene	Perylene	2×10^{11}
1-chloroanthracene	Rubrene	2×10^{11}
9-cyanoanthracene	Rubrene	3×10^{11}

solvent or the temperature of the system are changed. Obviously rate constants for a process controlled by diffusion would alter markedly if either the solvent viscosity or the temperature were changed.

4.2.2 Chemical reaction

A photo-excited molecule may undergo a chemical reaction and in doing so the excitation energy of the molecule will be dissipated, i.e. chemical

KINETICS OF PHOTOCHEMICAL PROCESSES

reaction provides an intermolecular route for the decay of an excited state species. If this is taken into account then the possible paths for the decay of an excited state species will be as given in Fig. 4.4. Analysis of a reaction scheme involving the processes of Fig. 4.4 can give, in some cases, a kinetic expression which enables the rate constant for the chemical reaction step to be evaluated. The principle of the method used for deriving the appropriate rate equations will be outlined with reference to a reaction scheme in which it is assumed that (a) the photo-excited molecule M^* reacts with a reactant R in a single bimolecular step:

$$M^* + R = \text{radicals or stable products}$$

and (b) the products of the bimolecular reaction do not react with either M or M^*.

The photo-excited molecule M^* is most likely to react while in the S_1 or T_1 state, and it is assumed for the present discussion that it is the triplet state molecules which react. The kinetic scheme for the photophysical and photochemical processes involving M will then be as follows:

Process	Representation	Rate
Excitation	$S_0(M) + h\nu \to S_1(M^*)$	I
Internal conversion	$S_1 \to S_0$	$k_{ic}[S_1]$
Fluorescence	$S_1 \to S_0 + h\nu_f$	$k_f[S_1]$
Intersystem crossing	$S_1 \to T_1$	$k_{isc}[S_1]$
Phosphorescence	$T_1 \to S_0 + h\nu_p$	$k_p[T_1]$
Intersystem crossing	$T_1 \to S_0$	$k'_{isc}[T_1]$
Chemical reaction	$T_1(M^*) + R \to$ radicals or stable products	$k_r[T_1][R]$
Energy transfer	$T_1(M^*) + S_0(A) \to S_0(M) + T_1(A)$	$k_e[T_1][A]$

where k_r represents the rate constant for the chemical reaction step.

83

INTRODUCTION TO MOLECULAR PHOTOCHEMISTRY

The last step in the above scheme represents energy transfer from the T_1 state of M to an acceptor species A which may be present as adventitious impurity or as an acceptor added to quench the T_1 state.

Applying the stationary state hypothesis to the T_1 state of M gives:

$$[T_1] = \frac{k_{isc}[S_1]}{k_p + k'_{isc} + k_r[R] + k_e[A]} \tag{4.49}$$

Now the rate of conversion of the molecule M into products is controlled by the rate of the chemical reaction step. This is equal to $k_r[T_1][R]$ and so the expression for the quantum yield for the consumption of M in the reaction is:

$$\phi = \frac{k_r[T_1][R]}{I} \tag{4.50}$$

Substituting for $[T_1]$ from Equation (4.49) into Equation (4.50) and rearranging gives:

$$\frac{1}{\phi} = \frac{I}{k_{isc}[S_1]}\left[1 + \frac{k_d}{k_r}\frac{1}{[R]} + \frac{k_e}{k_r}\frac{[A]}{[R]}\right] \tag{4.51}$$

where k_d is equal to $k_p + k'_{isc}$.

If quenchers are rigorously excluded from the reaction system, i.e. $[A] = 0$, Equation (4.51) then reduces to

$$\frac{1}{\phi^0} = \frac{I}{k_{isc}[S_1]}\left[1 + \frac{k_d}{k_r}\frac{1}{[R]}\right] \tag{4.52}$$

where ϕ^0 is the quantum yield for the consumption of M in the absence of quencher.

Dividing Equation (4.51) by Equation (4.52) gives the following expression for the ratio of the quantum yields for the consumption of M in the absence and presence of quencher:

$$\frac{\phi^0}{\phi} = 1 + \frac{k_e}{k_d + k_r[R]}[A] \tag{4.53}$$

Now the lifetime of the T_1 state of M is equal to the reciprocal of the sum of the rates of the various processes deactivating the state. In the absence of quencher the lifetime τ of the T_1 state is given by the expression

$$\tau = \frac{1}{k_d + k_r[R]} \tag{4.54}$$

An alternative expression is

$$\frac{1}{\tau} = k_d + k_r[R] \tag{4.55}$$

Combining Equations (4.53) and (4.54) gives the equation

$$\frac{\phi^0}{\phi} = 1 + k_e \tau [A] \tag{4.56}$$

Equations (4.55) and (4.56) can be used to derive a value for the rate constant, k_r, for the chemical reaction step in the following way. The ratio ϕ^0/ϕ is experimentally measured with known concentrations of A present and a fixed concentration of R. The value of the product $k_e \tau$ in Equation (4.56) is then obtained from the slope of the plot of ϕ^0/ϕ against A. The process is then repeated for different concentrations of R and a set of values for the product $k_e \tau$ is obtained. The energy acceptor used in such experiments is chosen so that the energy of its T_1 state is well below that of the T_1 state of the photo-excited reactant. Consequently the rate constant k_e

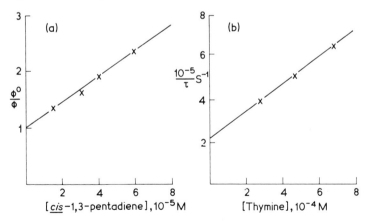

FIGURE 4.7
Photodimerization of thymine in acetonitrile, (a) plot for Equation (4.56), cis-1,3-pentadiene used as energy acceptor, concentration of thymine = $2 \cdot 7 \times 10^{-4}$ M, (b) plot for Equation (4.55). (Adapted from P. J. Wagner and D. J. Bucheck, *J. Amer. Chem. Soc.*, **92**, (1970) 181.)

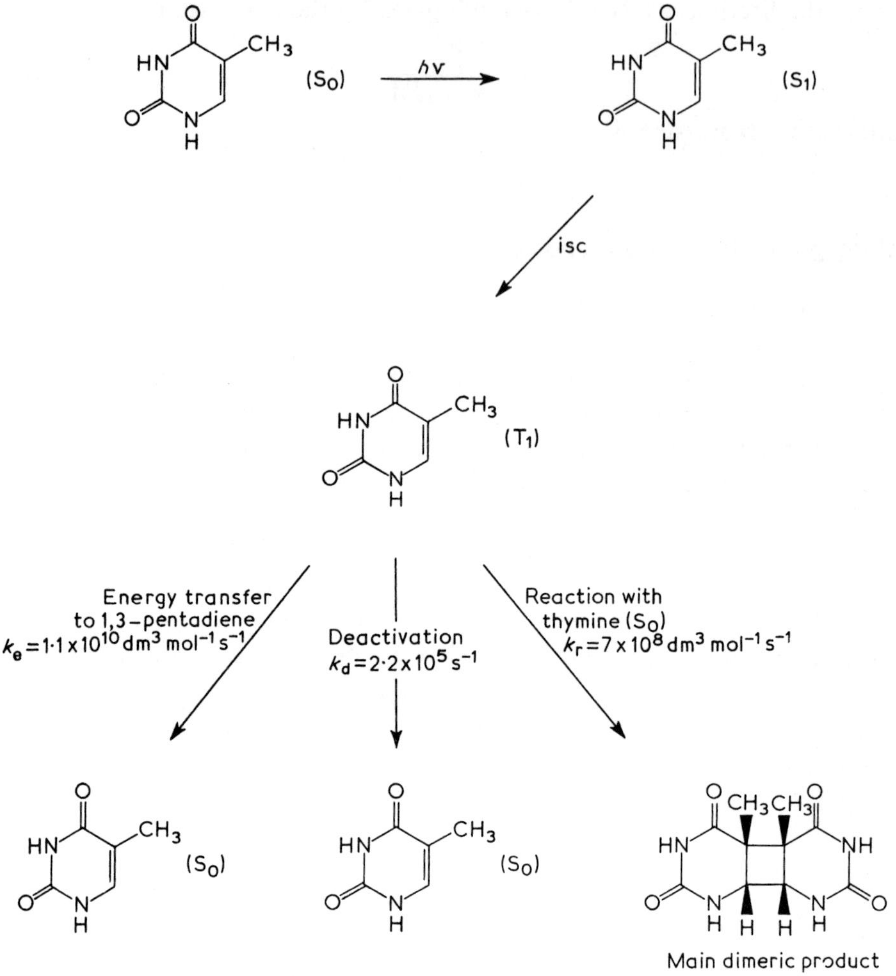

FIGURE 4.8
Mechanism for the photoreaction of thymine in acetonitrile. (Adapted from P. J. Wagner and D. J. Bucheck, *J. Amer. Chem. Soc.*, **92**, (1970) 181.)

for triplet–triplet energy transfer will equal the diffusion-controlled rate constant, a value for which can be calculated from Equation (3.7). Knowing k_e the values of τ can be calculated from the values obtained for the product $k_e\tau$. A plot corresponding to Equation (4.55) is then drawn up and the values of k_d and k_r can be obtained from the intercept and slope of this plot respectively.

The plots corresponding to Equations (4.55) and (4.56) are shown in Fig. 4.7 for the photolysis of thymine in acetonitrile. The chemical reaction in this system is the combination of triplet state thymine with ground state thymine to give a photodimer. The mechanism for the photo-reaction of thymine and the values obtained for k_d and k_r for this reaction are given in Fig. 4.8.

Photochemical reactions—I   5

Photo-excited molecules are highly energetic and have the potential for internal rearrangement or reaction with other molecular species in the system. The products of photochemical reactions frequently differ from those of thermal reactions and there are many examples where products obtained photochemically could not be prepared by any other synthetic route. The application of photochemistry to organic synthesis is rapidly increasing and the number of photochemical reactions now characterized is very large. As knowledge of photochemical principles develops so many of the apparently unique reactions are seen to be examples of general photochemical reaction processes. In this Chapter a number of general types of photochemical reaction are considered viz. photoreduction, photodimerization, photo-addition, photo-oxidation and photorearrangement.

5.1 Photoreduction

The photoreduction of carbonyl compounds in the presence of hydrogen donors has been one of the most extensively investigated photochemical reactions and the elucidation of the mechanism of this reaction has contributed greatly to the present-day understanding of photochemistry. In the great majority of cases the photoreduction occurs via the T_1 state of the photo-excited carbonyl compound. Molecules in this state react by abstracting hydrogen from a subtrate species RH to give a hydroxymethyl radical and a radical $R^{\bullet}$:

$$R'_2C=O\,(T_1) + RH \rightarrow R'_2\dot{C}OH + R^{\bullet}$$

The final reaction products are formed by self-combination and cross-combination reactions of the radicals produced in the hydrogen abstraction step. In the simplest case where the substrate RH is the alcohol derived from the ketone which is photolyzed, only one reaction product will be obtained. For example, the photoreduction of benzophenone by benzhydrol will give one radical species in the abstraction step and subsequent radical combination will give benzpinacol as the sole product:

$$(C_6H_5)_2C=O\ (T_1) + (C_6H_5)_2CHOH \rightarrow (C_6H_5)_2\dot{C}OH + (C_6H_5)_2\dot{C}OH$$

$$(C_6H_5)_2\dot{C}OH + (C_6H_5)_2\dot{C}OH \rightarrow (C_6H_5)_2\underset{OH}{\underset{|}{C}}-\underset{OH}{\underset{|}{C}}(C_6H_5)_2$$

The efficiency of the hydrogen abstraction step is dependent upon factors such as the nature of the T_1 state of the carbonyl compound, the structure of the carbonyl compound, the dissociation energy of the R—H bond and the type of solvent used.

Normally hydrogen abstraction by the carbonyl group is efficient when the T_1 state is (n, π^*) and inefficient when the T_1 state is (π, π^*). Thus, acetophenone, benzophenone and acetone with T_1 states of the (n, π^*) type are readily photoreduced in isopropanol whereas 1-naphthaldehyde, 2-acetonaphthone and p-phenylbenzophenone with T_1 states of the (π, π^*) type do not react. The difference in photochemical reactivity arises because of the difference between the electronic distribution in (n, π^*) and (π, π^*) excited states. An $n \rightarrow \pi^*$ electronic transition causes the promotion of an electron from a non-bonding orbital localized on the oxygen atom to an antibonding orbital delocalized over the C=O group, i.e. the transition effectively reduces the negative charge near the oxygen atom. When a $\pi \rightarrow \pi^*$ transition occurs an electron from a π orbital is promoted into an antibonding π^* orbital and since the π^* orbital is more closely associated with the oxygen atom than the π orbital there will be an increase in the negative charge near the oxygen atom. Thus the oxygen atom in the (n, π^*) excited state acts as an electrophile and hydrogen abstraction will be favoured, whereas in the (π, π^*) excited state the oxygen atom is effectively a nucleophile and hydrogen abstraction will be hindered.

Carbonyl compounds possessing T_1 states of the (π, π^*) type can be photoreduced, however, provided the dissociation energy of the R—H bond in the hydrogen donor is low. For example, 2-acetonaphthone is photo-reduced when tributylstannane is used as the hydrogen donor:

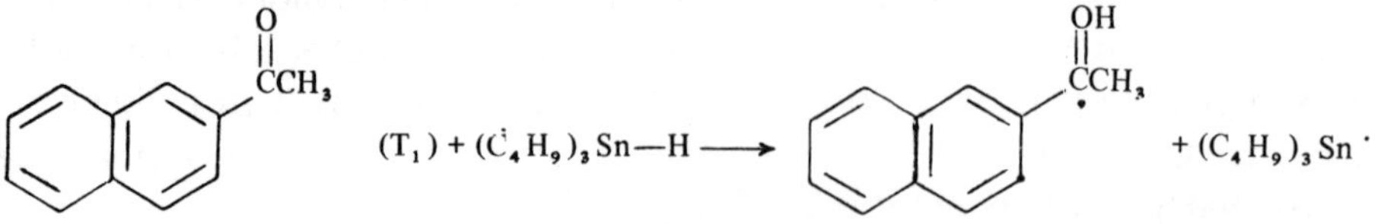

$(T_1) + (C_4H_9)_3\text{Sn}-\text{H} \longrightarrow$ [2-naphthyl C(OH)CH₃ radical] $+ (C_4H_9)_3\text{Sn}\cdot$

The T_1 state of certain *para* substituted benzophenones is neither (n, π^*) nor (π, π^*) but is instead a *charge-transfer* state. This situation can arise in compounds, such as *p*-aminobenzophenone, where the substituent is a strongly electron releasing group. Electronic excitation within such a compound can lead to a transfer of charge from the electron-donating group to the carbonyl group, and for the example of *p*-aminobenzophenone the structure of the resulting excited state would be

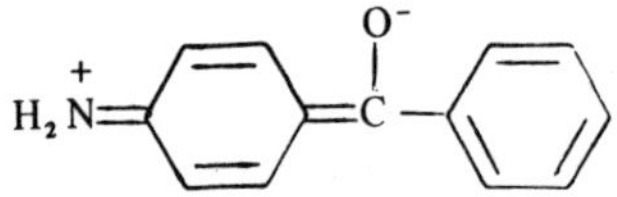

$H_2\overset{+}{N}=\text{C}_6H_4=\overset{O^-}{C}-C_6H_5$

Structures of the above type will be stabilized in polar solvents and in such solvents the charge-transfer state can be of lower energy than the potentially reactive (n, π^*) state. The negative charge on the oxygen atom in the charge-transfer state inhibits hydrogen abstraction and hence photo-reduction will be inefficient in polar solvents. In non-polar solvents the (n, π^*) state is normally of lower energy than the charge-transfer state and hydrogen abstraction occurs readily.

Ortho-substituted alkyl benzophenones can undergo *intra*molecular hydrogen abstraction provided they can form an intermediate six-membered transition state. This is shown below for the photoreaction of *o*-methylbenzophenone:

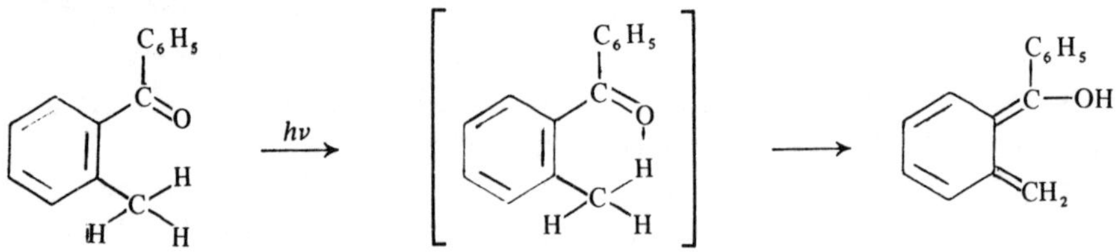

PHOTOCHEMICAL REACTIONS – I

The photo-enol produced as a result of the intramolecular hydrogen abstraction in o-methylbenzophenone is unstable but it can be trapped out of the reaction mixture as a Diels–Alder adduct with maleic anhydride.

*Ortho-tert*butylbenzophenones cannot form six-membered transition states such as shown above and these compounds do not photoreduce intramolecularly.

Intramolecular hydrogen abstraction occurs on photolysis of substituted valerophenones:

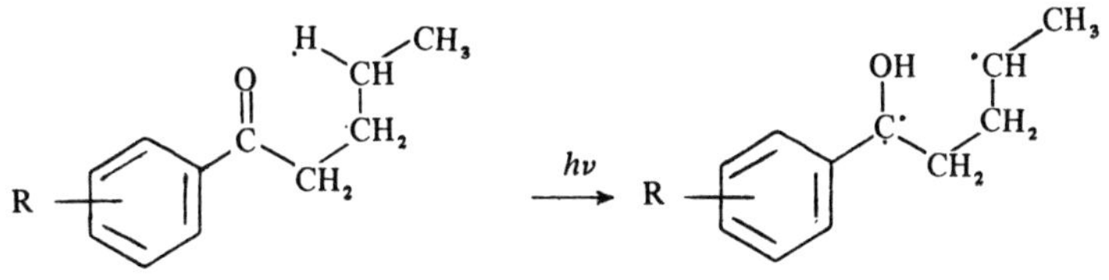

$$\text{Ar-CO-CH}_2\text{-CH}_2\text{-CH(CH}_3\text{)-H} \xrightarrow{h\nu} \text{Ar-C(OH)}\cdots\text{CH(CH}_3\text{)}\cdots$$

The above reaction takes place via the T_1 state of the valerophenone derivative. This state is principally (π, π^*) in character but because the energy of the T_1 state of derivatives of valerophenone is close to the energy of the T_2 (n, π^*) state there will be some 'mixing' of these states. This means that some (n, π^*) character will be imparted to the T_1 state. The extent of mixing of the states and thus the degree of (n, π^*) character of the T_1 state is dependent upon the nature and position of the substituent R. The greater the degree of (n, π^*) character in the T_1 state the more efficient is the hydrogen abstraction step.

A variety of hydrogen donors have been used for the photoreduction of carbonyl compounds. Included amongst them are amines, alcohols, hydrocarbons, phenols and amides. The reactions reported to date are far too numerous to itemize but some indication of the types of reaction undergone can be gained from the following list (see Bibliography for further examples):

(a) $(C_6H_5)_2C=O + RCH_2NHR' \xrightarrow{h\nu} (C_6H_5)_2\underset{\underset{OH}{|}}{C}-\underset{\underset{OH}{|}}{C}(C_6H_5)_2 + RCH=NR'$

INTRODUCTION TO MOLECULAR PHOTOCHEMISTRY

(b) 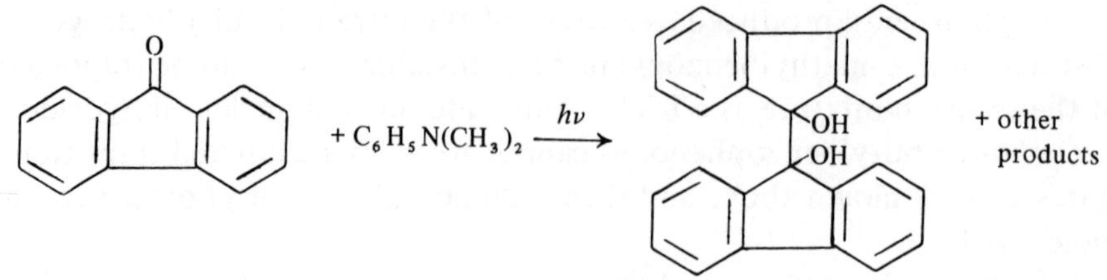 [fluorenone] + $C_6H_5N(CH_3)_2$ $\xrightarrow{h\nu}$ [bis-fluorenol structure with OH, OH] + other products

(c) [camphorquinone structure] + $(CH_3)_2CHOH$ $\xrightarrow{h\nu}$ [product with OH, H] + [product with OH, H]

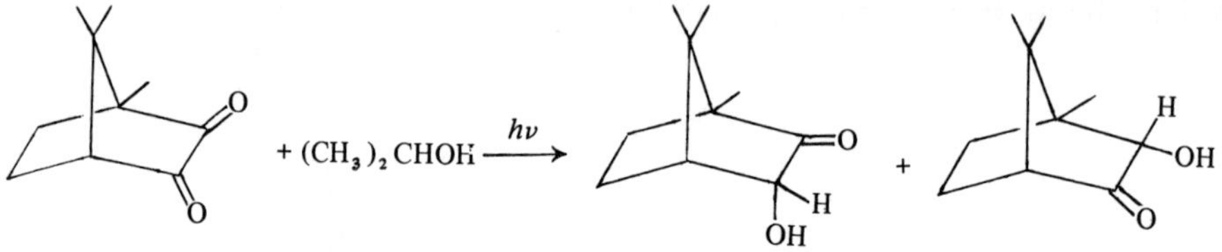

(d) 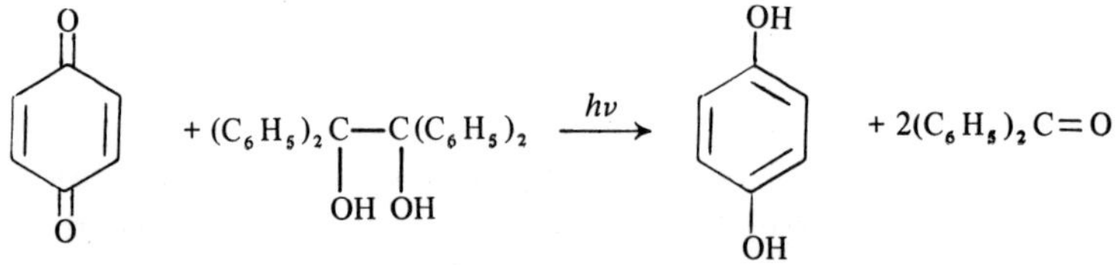 [benzoquinone] + $(C_6H_5)_2C\!-\!C(C_6H_5)_2$ with OH OH $\xrightarrow{h\nu}$ [hydroquinone with OH, OH] + $2(C_6H_5)_2C=O$

(e) $(C_6H_5)_2C=O + CH_3CON(CH_3)_2$ $\xrightarrow{h\nu}$ $(C_6H_5)_2\underset{\underset{\text{OH}}{|}}{C}CH_2CON(CH_3)_2$

The photoreduction of aryl ketones by amines is of particular interest in that recent studies have shown that a charge-transfer interaction occurs between the triplet state ketone and the amine prior to the transfer of hydrogen from the amine to the ketone. This is shown in the following scheme:

$$Ar_2C=O\ (T_1) + RCH_2NR'_2 \rightarrow [Ar_2\dot{C}\!-\!\bar{O} \quad RCH_2\overset{+}{\dot{N}}R_2]$$

(a) ↙   ↘ (b)

$Ar_2C=O + RCH_2NR_2$ $Ar_2\dot{C}\!-\!OH + R\dot{C}HNR_2$

The amine acts by donating an electron to the triplet state ketone; a ketone radical anion and an amine radical cation are formed. These radicals interact either quenching the photoreduction (process (a)) or initiating the

92

hydrogen transfer (process (b)). The involvement of a charge-transfer step in the photoreduction is supported by the fact that the reactivity of amines increases as their ionization potentials decrease. That is, the more easily an amine can lose an electron the more effective it is as a reducing agent in the photoreduction.

5.2 Photodimerization

Photodimerization involves the combination of an electronically excited molecule with a *like* ground state molecule to give a 1:1 photo-adduct. Olefinic compounds have a tendency to dimerize when irradiated and many examples are known of the dimerization of aromatic hydrocarbons, mono-olefins, conjugated dienes, and α, β-unsaturated carbonyl compounds. Photodimerization of olefinic compounds can occur by either (a) 1,2–1,2 addition, (b) 1,2–1,4 addition, or (c) 1,4–1,4 addition. Examples of these addition processes are shown below:

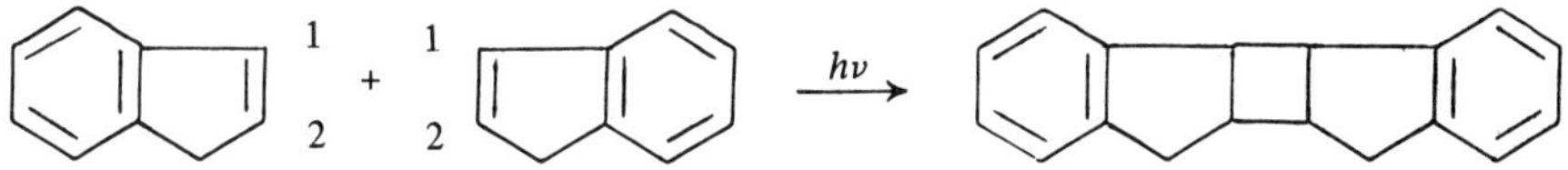

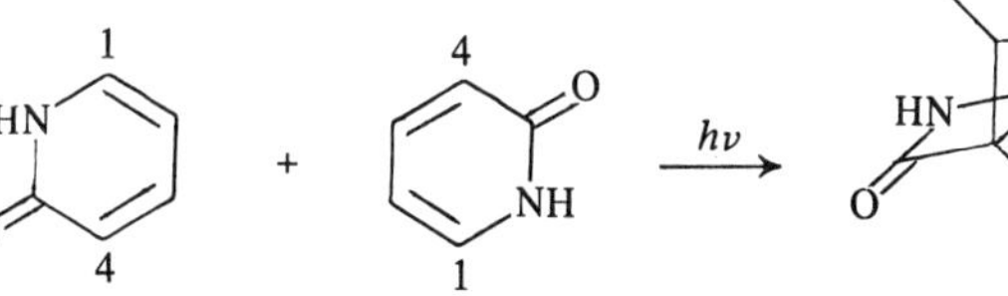

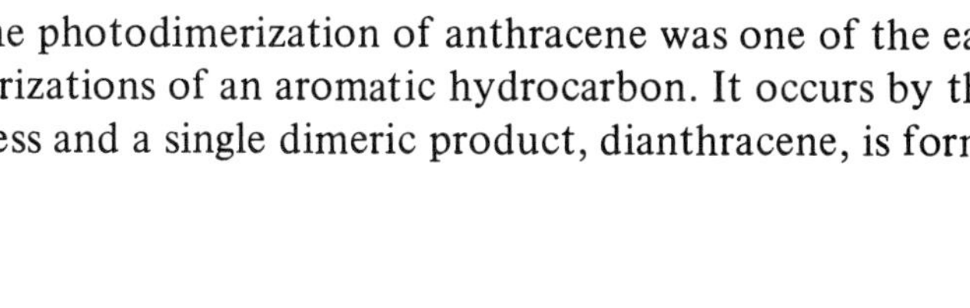

The photodimerization of anthracene was one of the earliest reported dimerizations of an aromatic hydrocarbon. It occurs by the 1,4–1,4 addition process and a single dimeric product, dianthracene, is formed (see below).

INTRODUCTION TO MOLECULAR PHOTOCHEMISTRY

The dimerization is unaffected by the presence of oxygen and since oxygen is an efficient quencher of triplet states but not of singlet states it is concluded that the S_1 (π, π^*) state of anthracene is the photoreactive state. The dianthracene is formed by attack of singlet-excited anthracene on a ground state anthracene molecule:

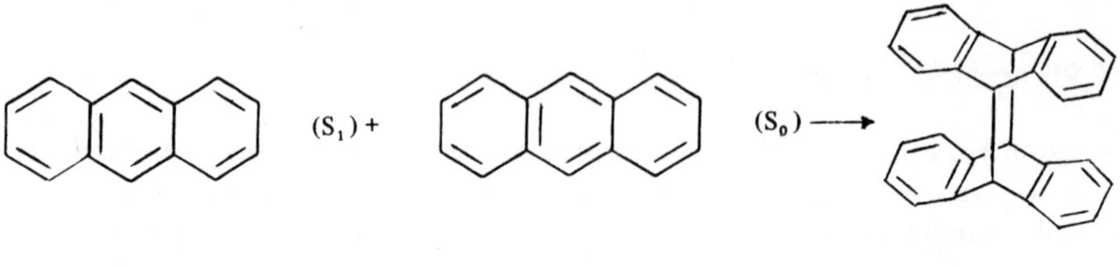

Several 9-substituted anthracenes, such as 9-anthraldehyde and 9-methylanthracene, have been dimerized successfully but 9,10-disubstituted anthracenes have not been dimerized because of steric interaction.

Acenaphthylene dimerizes via 1,2–1,2 addition with the formation of the two stereoisomers (A) and (B) below:

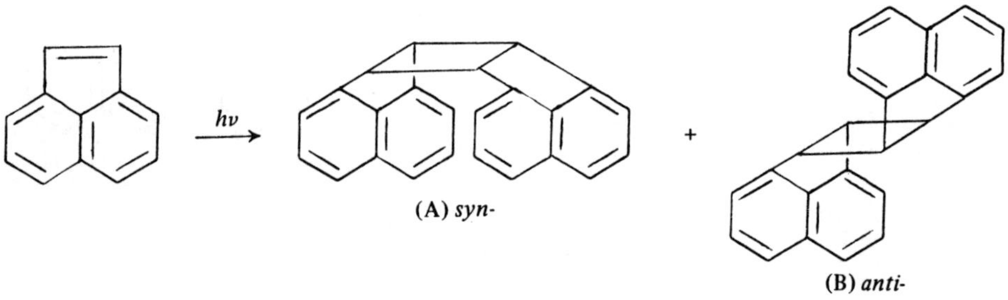

An interesting feature of the acenaphthylene dimerization is that the *syn-*dimer (A) is formed via a triplet state while the *anti-* dimer (B) is formed via a singlet state. Evidence for this is that triplet state quenchers inhibit the formation of (A) relative to (B) while heavy atom solvents, which promote $S_1 \leadsto T_1$ intersystem crossing, favour the formation of (A) relative to (B).

The photo-induced dimerizations of butadiene and isoprene proceed principally via 1,2–1,2 addition to give cyclobutane derivatives as the main products. However, 1,2–1,4 and 1,4–1,4 additions occur to some extent and cyclohexene and cyclooctadiene derivatives are also present in the product mixtures. The product ratios for the acetophenone-sensitized dimerization of butadiene and isoprene are as shown below:

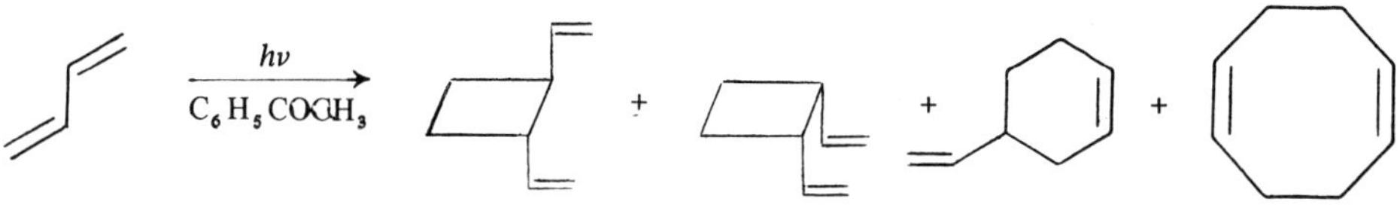

The photodimerizations of butadiene and isoprene proceed via the T_1 (π, π^*) state of the olefin. This state can be populated either by direct irradiation of the olefin or by triplet energy transfer from sensitizers such as acetophenone, benzophenone and anthraquinone. It is experimentally more convenient to induce dimerization using sensitizers since the sensitizers absorb at relatively long wavelengths as compared to butadiene and isoprene. For example, the longest wavelength absorption in benzophenone lies at 270 nm whereas that for butadiene and isoprene is at 217 and 220 nm respectively.

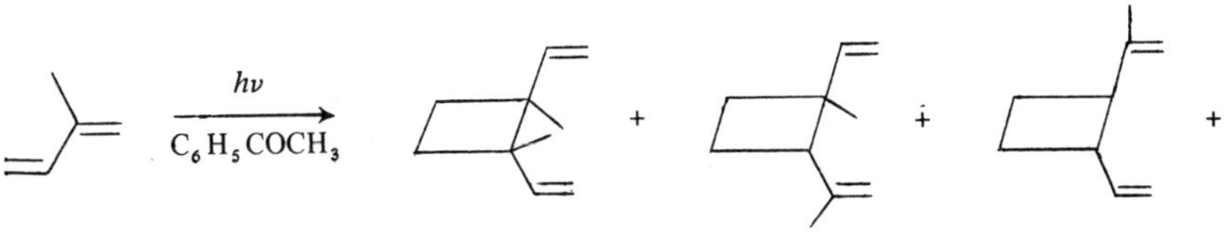

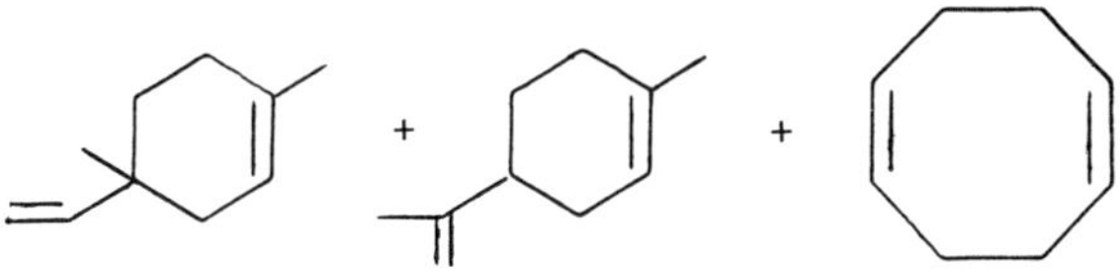

The photodimerization of cyclopentadiene provides a further example of the dimerization of a conjugated diene. In this case a triplet state sensitizer is required in the system to ensure efficient reaction:

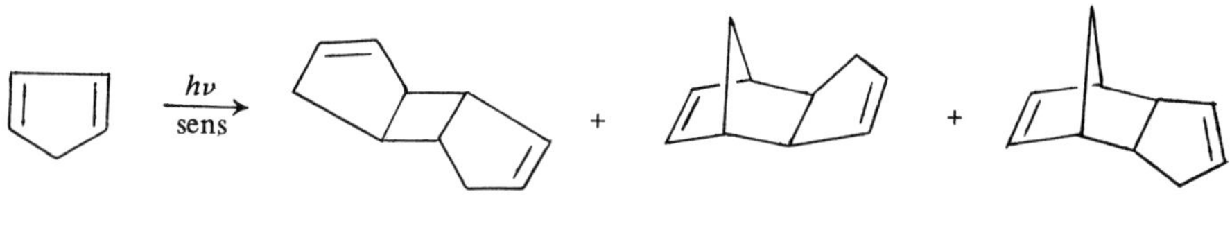

INTRODUCTION TO MOLECULAR PHOTOCHEMISTRY

Whereas aromatic hydrocarbons and conjugated dienes dimerize via a (π, π^*) excited state the dimerization of α, β-unsaturated carbonyl compounds can occur from either an (n, π^*) or a (π, π^*) state. This state can be either singlet or triplet.

Coumarin provides an interesting example of dimerization of an α, β-unsaturated carbonyl compound. The two main products formed are shown below:

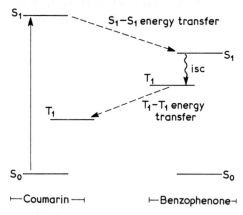

(A) *syn-*   (B) *anti-*

Dimer (A) is formed via singlet state coumarin and is the major product when coumarin is irradiated in the absence of sensitizers. Dimer (B) is formed via triplet state coumarin and is the major product when coumarin is irradiated in the presence of benzophenone. The mechanism for the formation of triplet state coumarin in the presence of benzophenone is shown in Fig. 5.1. In this rather remarkable system the excitation energy is transferred from coumarin to benzophenone and then transferred back again.

FIGURE 5.1
Mechanism for the population of the triplet state of coumarin in the presence of benzophenone.

Other α, β-unsaturated carbonyl compounds which have been observed to undergo photodimerization include maleic anhydride, 1,4-naphthaquinone and cyclohexenone:

(a) 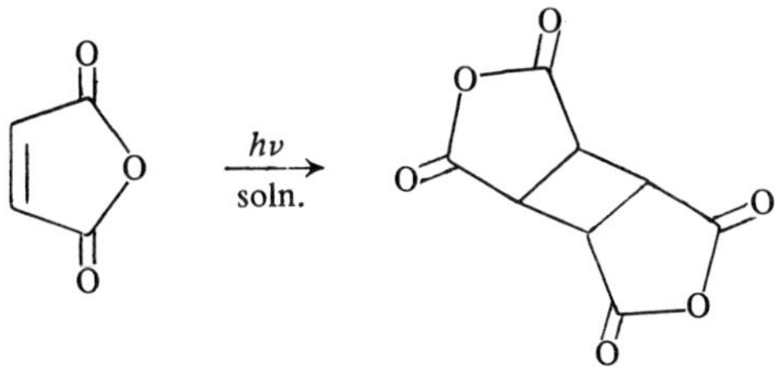 [maleic anhydride] →(hν, soln.) [dimer]

(b) 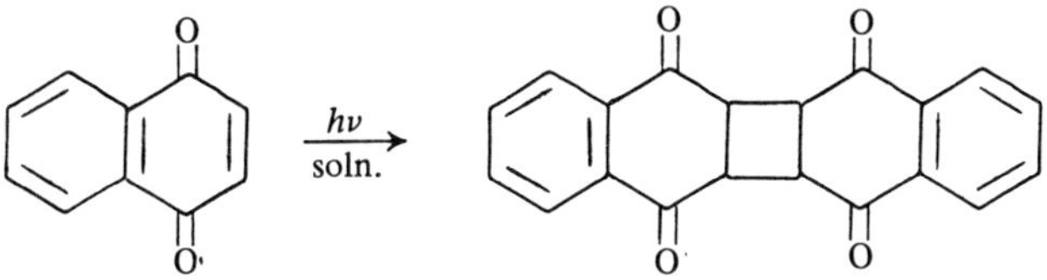 [1,4-naphthaquinone] →(hν, soln.) [dimer]

(c) 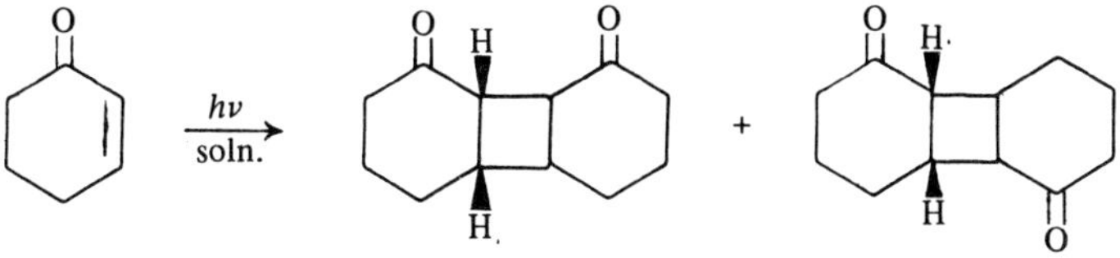 [cyclohexenone] →(hν, soln.) [cis-fused dimer] + [trans-fused dimer]

5.3 Photo-addition

Photo-addition may be defined as the combination of an electronically excited molecule with an *unlike* ground state molecule to give a 1 : 1 photo-adduct. Such combination reactions are commonly encountered when mixtures of olefins are irradiated. The number of possible combinations is very large and mixed bimolecular additions involving quinones, aldehydes, ketones, olefins and aromatic hydrocarbons are known. Examples of these reactions are considered in this section, with particular reference being paid to photo-addition to carbonyl and to aromatic compounds.

5.3.1 *Photo-addition to the carbonyl group*

Photo-addition frequently results in the linking of two addends via a four-membered ring system. In the case of carbonyl-olefin systems such photo-additions give rise to the formation of an oxetan structure:

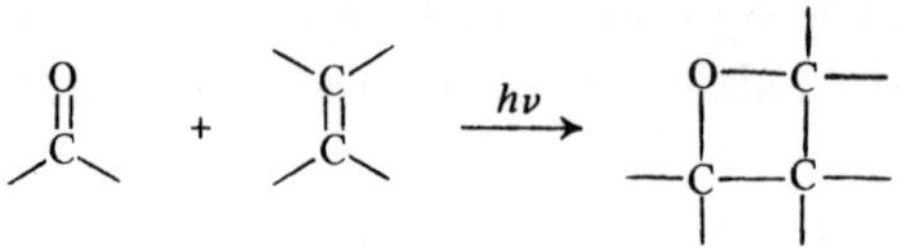

INTRODUCTION TO MOLECULAR PHOTOCHEMISTRY

Some examples of carbonyl-olefin systems which give oxetans on irradiation are shown below:

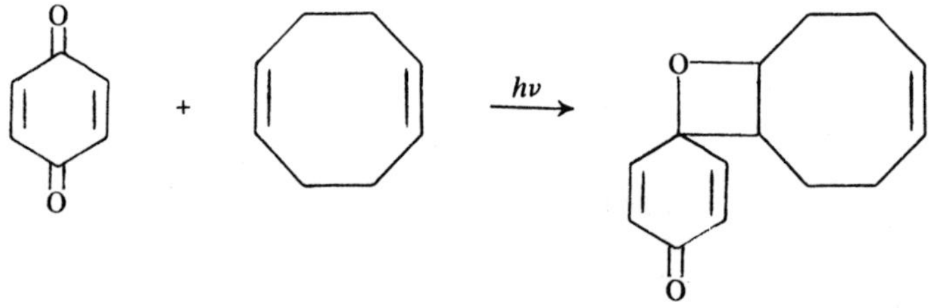

(a) 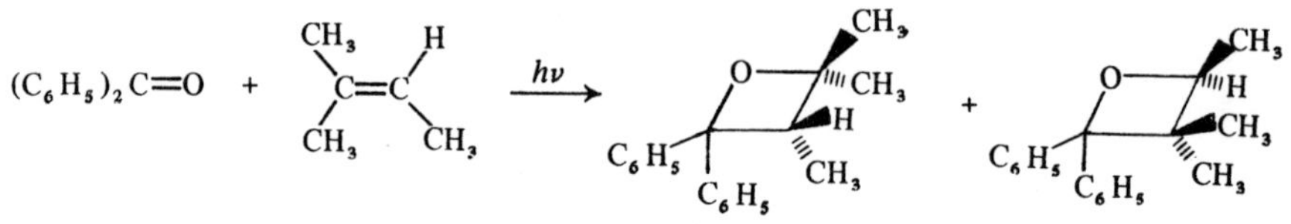 $(C_6H_5)_2C=O$ + (2-methyl-2-butene) $\xrightarrow{h\nu}$ oxetane isomers

(b) (benzoquinone) + (cyclooctatetraene) $\xrightarrow{h\nu}$ adduct

(c) CH_3CH_2CHO + (vinyl propyl ether, methyl-substituted) $\xrightarrow{h\nu}$ oxetane isomers

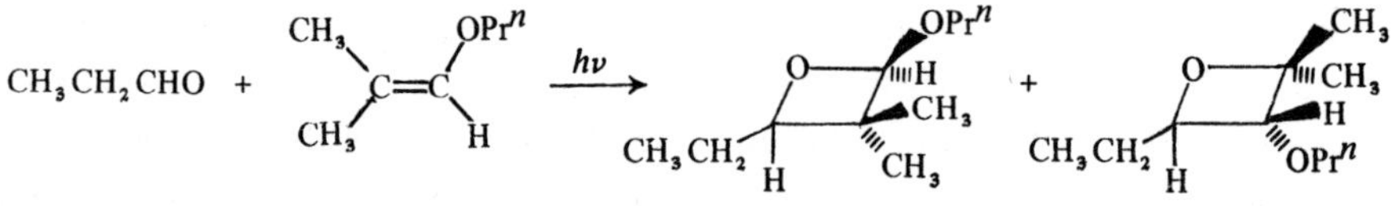

The cyclo-addition reaction is normally a two-step process involving the formation of an intermediate biradical or biradicals which can subsequently ring close to form oxetans. In reactions such as (a) and (c) above the relative amounts of the different oxetans formed in the reaction mixture is determined by the relative stabilities of the different biradical intermediates. This can be considered in relation to reaction (a) where there are four possible biradical intermediates:

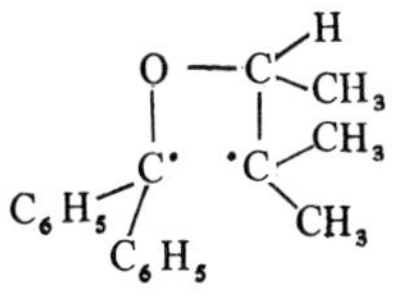

(A), 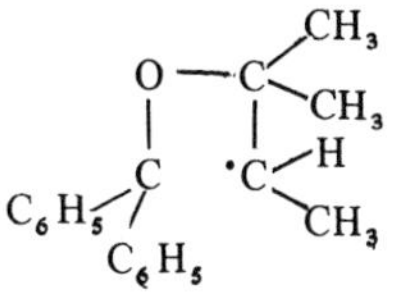 (B), 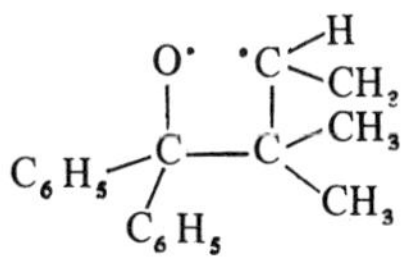 (C), 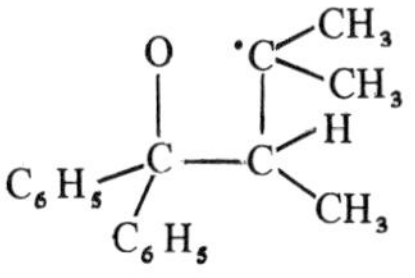 (D) — structural diagrams of biradical intermediates.

Intermediate (A) is the most stable since here the unpaired electrons are located at sites where they are conjugated with the maximum number of phenyl and methyl groups. The oxetan derived by ring closure in A constitutes 90% of the reaction product. Intermediate (B) is the next most stable in terms of electron conjugation with methyl and phenyl groups and the oxetan derived from it constitutes the remainder of the reaction product.

The cyclo-addition of aromatic carbonyl compounds to olefins usually involves the T_1 state of the carbonyl compound and, like hydrogen abstraction reactions, occurs most readily when the T_1 state is (n, π^*) in character. The initial step in photocyclo-additions is electrophilic attack of the photo-excited carbonyl on the olefin and as there is a greater degree of electrophilic character on the oxygen atom in the (n, π^*) state than in the (π, π^*) state those compounds with T_1 (n, π^*) states are more efficient reactants. Aromatic carbonyls with T_1 states of the (π, π^*) type can also add to olefins to form oxetans but the quantum yields of such reactions are generally low.

The efficiency of cyclo-addition of an aromatic carbonyl compound to an olefin is dependent upon the relative energies of the lowest triplet states of the two reactants. Oxetan formation is favoured when the triplet level of carbonyl compound is below that of the olefin but is suppressed when the triplet level is above that of the olefin. In the latter situation triplet–triplet energy transfer from the carbonyl compound to the olefin deactivates the carbonyl compound and prevents oxetan formation. The reaction of

INTRODUCTION TO MOLECULAR PHOTOCHEMISTRY

norbornene with acetophenone (E_{T_1} = 308 kJ mol^{-1}) and with benzophenone (E_{T_1} = 287 kJ mol^{-1}) provides an example where the course of the reaction is dependent upon the relative energies of the T_1 states of the reactants. The products formed in these systems are indicated below:

$$\text{norbornene} \xrightarrow{h\nu, \text{ acetophenone}} \text{norbornene dimers} + \text{no oxetan}$$

$$\text{norbornene} \xrightarrow{h\nu, \text{ benzophenone}} \text{no norbornene dimers} + \text{oxetan (80\%)}$$

The observations recorded in the above diagram indicate that the energy of the T_1 state of norbornene lies below that of acetophenone and above that of benzophenone. In the norbornene-acetophenone system the excitation energy taken up by acetophenone is transferred to norbornene and the triplet state norbornene so formed undergoes photodimerization. In the norbornene-benzophenone system energy transfer from benzophenone to norbornene does not occur and the excitation energy is utilized to form the oxetan product.

The formation of oxetans by reaction of alkyl aldehydes and alkyl ketones with olefins can apparently take place via either the singlet or triplet state of the carbonyl compound. The singlet state reaction is a one-step process as indicated by the fact that the stereo-arrangement of the substituent groups in the olefin is retained to some degree in the oxetan product. If the reaction were a two-step process involving the formation of an intermediate biradical then because of bond rotation in the intermediate any specific stereo-arrangement in the reactant would not be carried over to the product. The reaction of S_1 state acetone with *trans*-1,2-dicyanoethylene provides an example where the stereo-arrangement of substituent groups in the reactant is retained in the product. In this case the cyano groups are *trans* in the olefin and *trans* in the oxetan.

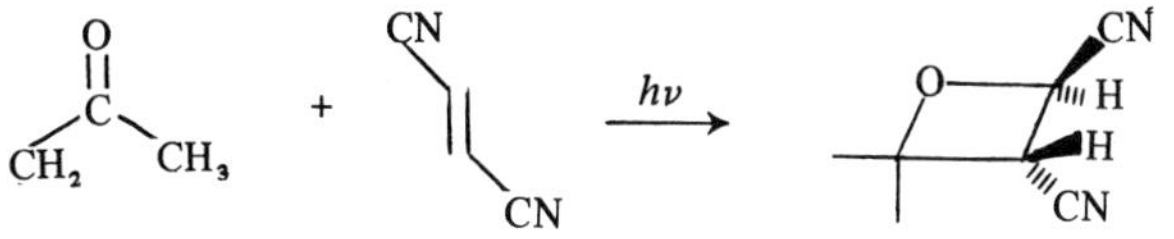

The photo-addition of acetone to cyclopentadiene apparently also proceeds via S_1 state acetone. The photo-addition is not likely to occur via the T_1 state of acetone since this state is of higher energy than the T_1 state of cyclopentadiene and would be deactivated by triplet–triplet energy transfer. The S_1 state, however, is of lower energy than the S_1 state of cyclopentadiene and hence would not be deactivated by singlet–singlet energy transfer. The reaction is illustrated below:

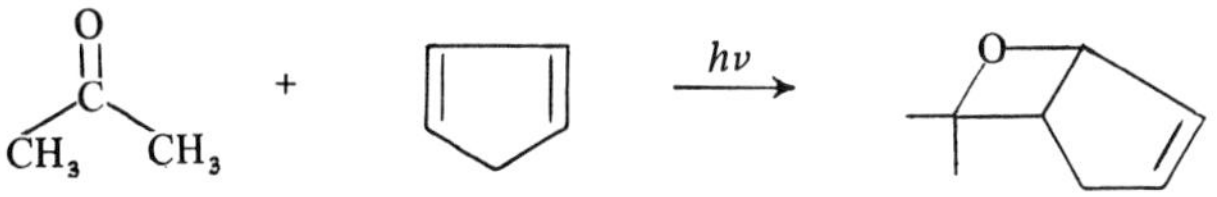

When both the lowest excited singlet and triplet states of acetone are of lower energy than the respective singlet and triplet states of the reactant olefin then the oxetan products may result from reaction of either of the excited states of acetone. This situation arises in the photo-addition of acetone to *cis* and *trans* 1-methoxy-1-butene. The main difference here between the singlet and triplet state reactions, in terms of the products, is that the singlet state reaction is rather more stereospecific. The stereospecificity of the singlet state reaction is seen in the photo-addition to *cis*-1-methoxy-1-butene where the yield of the *cis*-3-methoxyoxetan product is four times greater than the yield of the *trans*-3-methoxyoxetan.

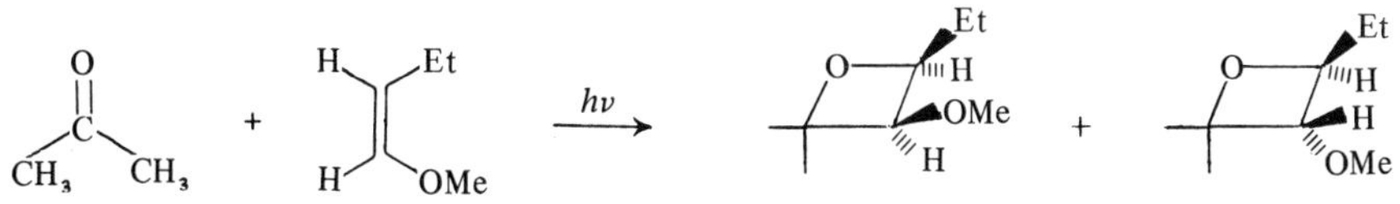

Carbonyl compounds also add to acetylenes on irradiation and it is thought that the reaction takes place via an oxeten intermediate:

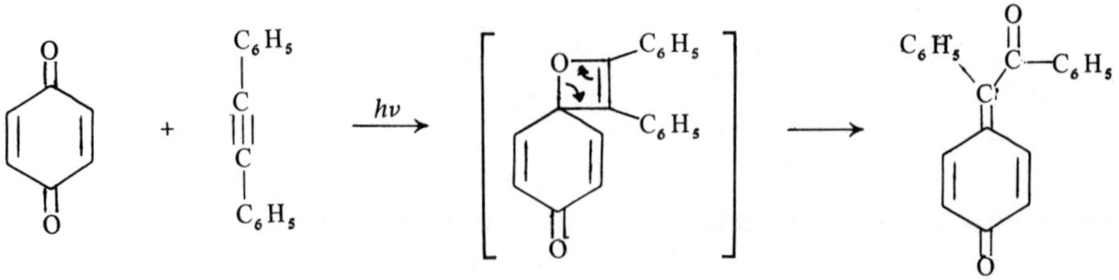

5.3.2 Photo-addition to aromatic compounds

Photo-addition to the benzene ring can occur by either 1,2, 1,3, or 1,4 addition. 1,2 additions are the most frequent and include the following reactions:

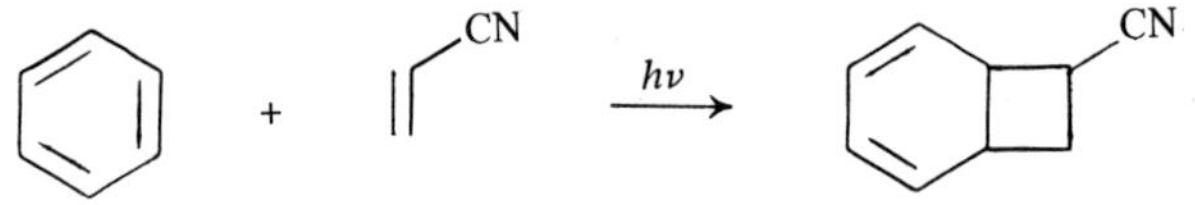

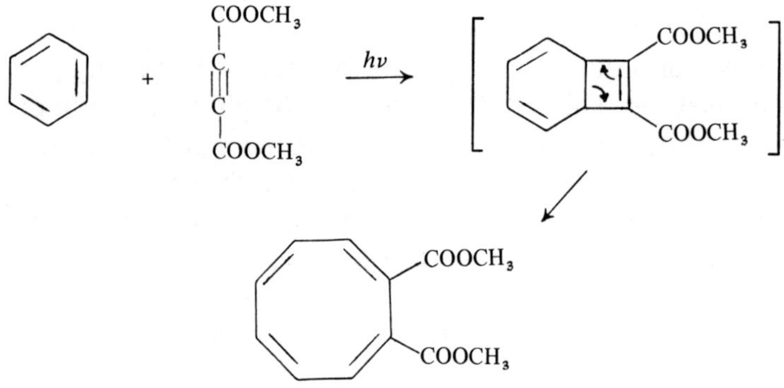

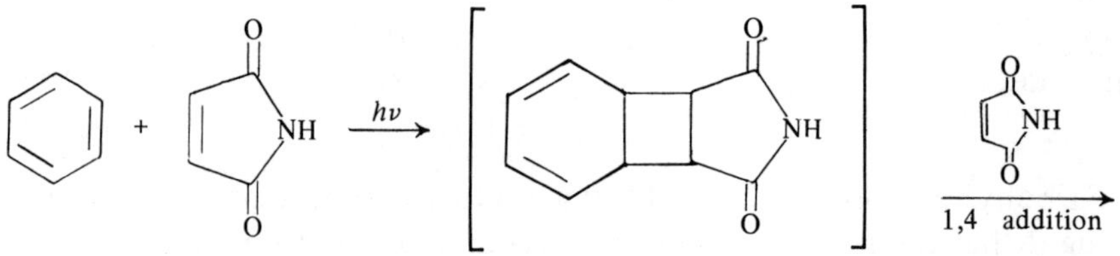

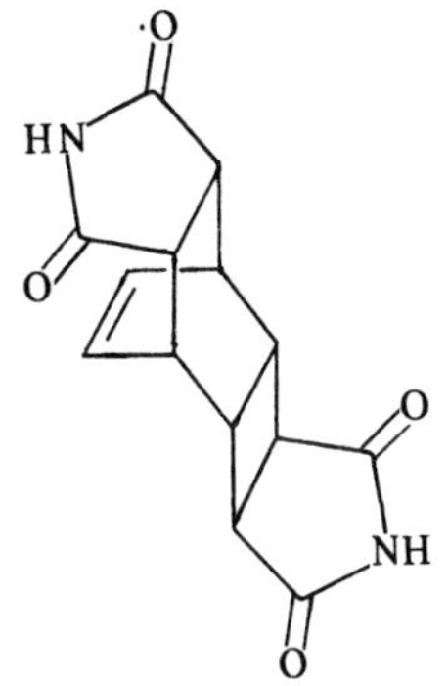

Reaction (c) above between benzene and maleimide is unlike the other photo-additions shown in that a 1 : 2 adduct is formed. However the photochemical step in this reaction is the same as in the other reactions in that a 1,2 addition of the olefinic compound occurs initially, and this is followed by a *thermal* 1,4 addition of a second olefinic molecule. It has been found that the benzene-maleimide adduct is formed both in the presence and in the absence of triplet state sensitizers and that oxygen inhibits adduct formation in the sensitized reactions but not in the unsensitized reactions. These observations indicate that the photo-chemical reaction takes place from both singlet and triplet states. Benzene and maleimide form an electron donor-electron acceptor complex in solution and it is thought likely that the reaction can be initiated from either the singlet or triplet state of this complex.

The photo-addition of acidified methanol to benzene gives products whose structures can be rationalized in terms of a 1,3 addition. The mechanism for the reaction involves conversion of S_1 state benzene to a biradical intermediate which forms an addition product with ground state methanol:

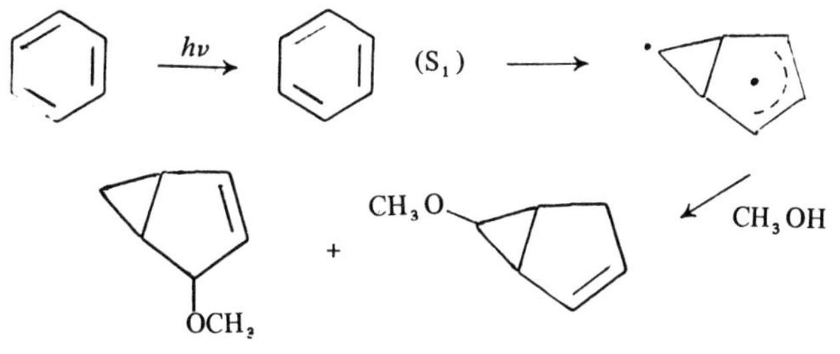

INTRODUCTION TO MOLECULAR PHOTOCHEMISTRY

The 1,4 addition of simple primary and secondary amines to benzene also apparently results from conversion of an excited state benzene species into a biradical intermediate. However this time the intermediate is formed from the T_1 state of benzene. The postulated mechanism for the photo-addition of benzene to pyrrole is as follows:

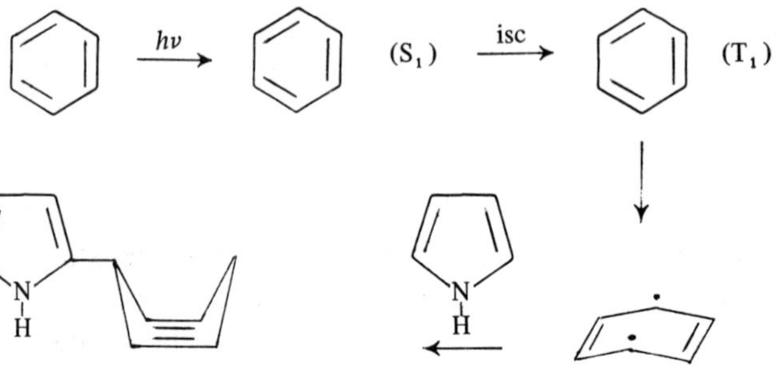

Maleic anhydride readily adds to aromatic hydrocarbons and a number of adducts have been isolated. Whereas a 1:2 adduct is formed with the monocyclic system in benzene only 1:1 adducts are formed with polycyclic aromatic hydrocarbons. The reaction with benzene is a two-step process involving consecutive 1,2 and 1,4 additions while the reaction with polycyclic aromatics is a single stage process involving either 1,2 or 1,4 addition. 1,2 additions is favoured in the latter systems when the site of attack has a high degree of olefinic character (see reaction (b) below). Examples of the photo-addition of maleic anhydride (MA) to aromatic systems are shown below:

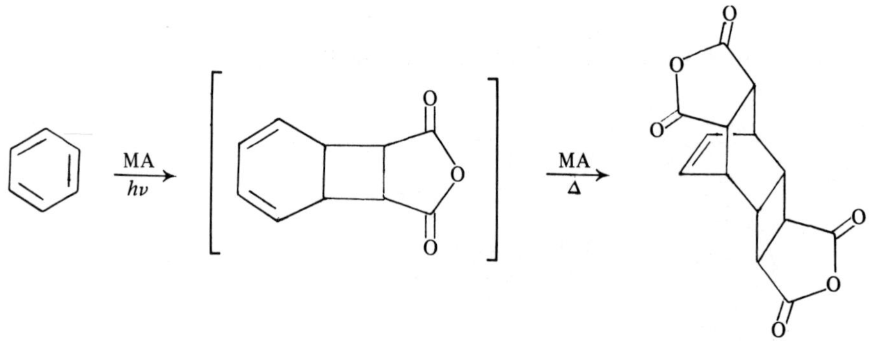

(b) 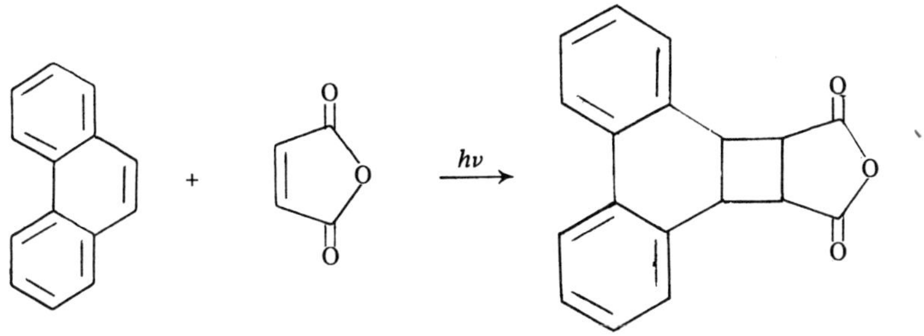 [reaction scheme: phenanthrene + maleic anhydride $\xrightarrow{h\nu}$ [2+2] cycloadduct]

(c) 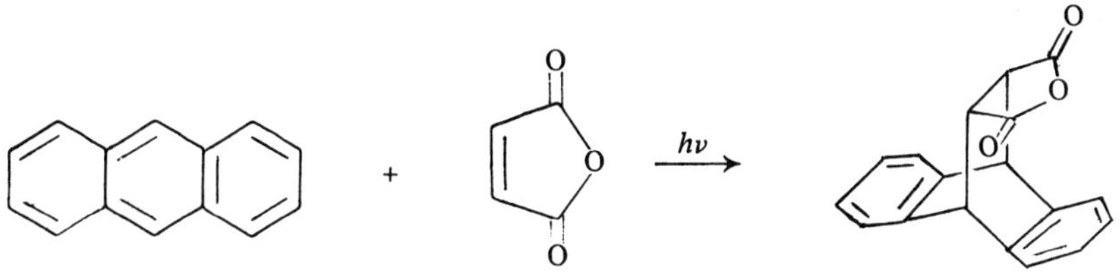 [reaction scheme: anthracene + furanone $\xrightarrow{h\nu}$ cycloadduct]

5.4 Photo-oxidation

Irradiation of organic compounds in the presence of a sensitizer and oxygen can lead to the involvement of oxygen in the photoreaction and the formation of oxygenated products. The presence of a sensitizer is necessary to the reaction and normally the photosensitized oxidations proceed via the T_1 state of the sensitizer. The sensitizer acts in one of two ways – either by abstracting hydrogen from the substrate to form radicals which subsequently react with the oxygen (Type I process) or by activating the oxygen so that a direct reaction can occur between oxygen and the substrate (Type II process).

The generalized mechanism for the Type I process is as follows:

$$\text{sens} + h\nu \rightarrow \text{sens}(S_1) \rightarrow \text{sens}(T_1)$$

$$\text{sens}(T_1) + RH \rightarrow {}^{\bullet}\text{sens-H} + R^{\bullet}$$

$$R^{\bullet} + O_2 \rightarrow RO_2^{\bullet}$$

$$RO_2^{\bullet} + RH \rightarrow ROOH + R^{\bullet}$$

$$RO_2^{\bullet} + {}^{\bullet}\text{sens-H} \rightarrow ROOH + \text{sens}$$

Examples of the Type I photo-oxidation are seen in the benzophenone-sensitized oxidations of secondary alcohols to hydroxy-hydroperoxides. The mechanism for the oxidation of isopropanol is shown below:

$$(C_6H_5)_2C=O\ (T_1) + (CH_3)_2CHOH \rightarrow (C_6H_5)_2\dot{C}OH + (CH_3)_2\dot{C}OH$$

$$(CH_3)_2\dot{C}OH + O_2 \rightarrow (CH_3)_2\overset{O-O^\bullet}{\underset{|}{C}}OH$$

$$(CH_3)_2\overset{O-O^\bullet}{\underset{|}{C}}OH + (CH_3)_2CHOH \rightarrow (CH_3)_2\overset{OOH}{\underset{|}{C}}OH + (CH_3)_2\dot{C}OH$$

$$(C_6H_5)_2\dot{C}OH + (CH_3)_2\dot{C}OH \rightarrow (C_6H_5)_2C=O + (CH_3)_2CHOH$$

It can be seen from the last step shown above that ground state benzophenone is regenerated in the course of the reaction. Thus benzophenone is acting as a true sensitizer since it is not consumed in the reaction.

In Type II photo-oxidations the reaction between oxygen and the organic substrate may occur either when (a) the oxygen is complexed with the sensitizer or when (b) the oxygen is in the S_1 state formed by energy transfer from the sensitizer. The two possible mechanisms for the Type II photo-oxidations are given below:

Type II(a)   sens + $h\nu$ → sens (S_1) → sens (T_1)

sens (T_1) + O_2 → •sens—O—O•

•sens—O—O• + M → sens + MO_2

Type II(b)   sens + $h\nu$ → sens (S_1) → sens (T_1)

sens (T_1) + O_2 → sens + O_2 (S_1)

O_2 (S_1) + M → MO_2

There is some evidence that Type II photo-oxidations in fact take place by the second of the two mechanisms above, although the first mechanism cannot be ruled out for all oxidation systems. The ground state of oxygen is a triplet state, T_0, and according to the spin conservation rules the energy transfer step in the Type II(b) process must produce singlet-excited state oxygen, i.e. S_1 state oxygen. The oxidation then takes place by the direct addition of singlet state oxygen to the organic substrate.

Dyes such as Rose Bengal and fluorescein are commonly used as the sensitizing agents in photo-oxidation reactions. Included among the organic compounds which can be oxidized using dye sensitizers are olefins, dienes, furans and polycyclic aromatic hydrocarbons. The oxidations, which occur by a Type II mechanism, are typified by the following examples:

(a) [cyclohexadiene] $\xrightarrow[\text{sens}]{h\nu, O_2}$ [endoperoxide]

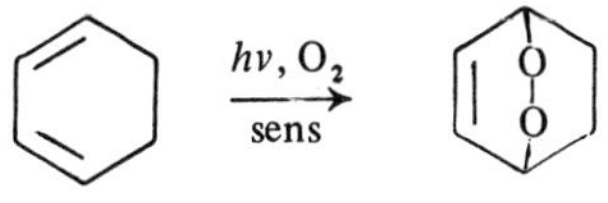

(b) [2,5-dimethylfuran] $\xrightarrow[\text{sens}]{h\nu, O_2}$ [ozonide product with CH_3 groups]

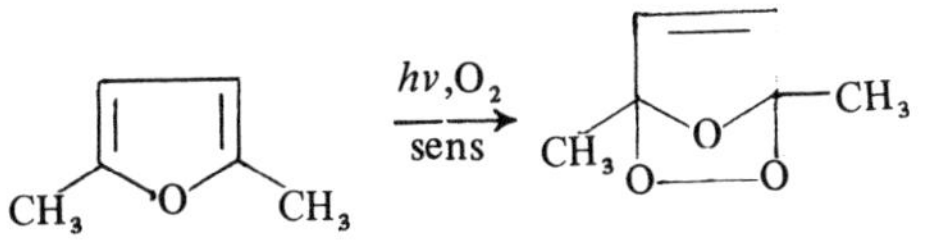

(c) [anthracene] $\xrightarrow[\text{sens}]{h\nu, O_2}$ [anthracene endoperoxide]

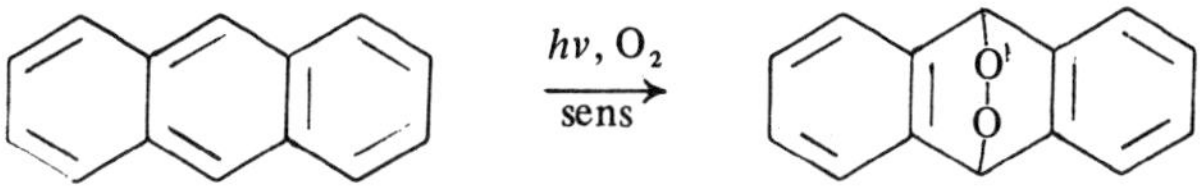

5.5 Photorearrangement

Electronically excited molecules have considerable excess energy over ground state molecules and it is not surprising that changes in the bonding system and of the positions of atoms in the molecular framework can occur in these energy rich species. Such rearrangements can lead to ground state products which are isomeric with the initial reactant. The reactant and products can be either *structural* isomers where groups or atoms in the isomers have entirely different positions or *valence-bond* isomers where the bonding system in the isomers differs and some of the atoms may have different relative positions. Examples of photo-induced structural and valence-bond isomerizations are considered in the following sections.

5.5.1 cis-trans isomerization

Molecular rearrangement by rotation about a double bond can be induced either by direct absorption of radiation or by energy transfer from a sensitizer molecule. Examples of sensitized and unsensitized photo-induced rotations about the C=C and N=N bonds are listed below:

(a) $\quad \underset{H}{\overset{C_6H_5}{>}}C=C\underset{C_6H_5}{\overset{H}{<}} \quad \xrightarrow[\text{sens}]{h\nu} \quad \underset{C_6H_5}{\overset{H}{>}}C=C\underset{C_6H_5}{\overset{H}{<}}$

(b) $\quad \underset{H}{\overset{HOOC}{>}}C=C\underset{COOH}{\overset{H}{<}} \quad \xrightarrow{h\nu} \quad \underset{HOOC}{\overset{H}{>}}C=C\underset{COOH}{\overset{H}{<}}$

(c) $\quad \underset{H}{\overset{C_6H_5}{>}}N=N\underset{C_6H_5}{\overset{H}{<}} \quad \xrightarrow{h\nu} \quad \underset{H}{\overset{C_6H_5}{>}}N=N\underset{H}{\overset{C_6H_5}{<}}$

(d) $\quad \underset{O}{\overset{R}{>}}N=N\underset{R}{\overset{O}{<}} \quad \xrightarrow{h\nu} \quad \underset{O}{\overset{R}{>}}N=N\underset{O}{\overset{R}{<}}$

The photosensitized isomerizations of stilbenes, 1,2-diphenylpropenes and 1,3-pentadienes have been studied in detail and it has been found that in each case both the *cis* → *trans* and the *trans* → *cis* photosensitized isomerizations proceed via a common (π, π^*) triplet state. This triplet state is often referred to as the perpendicular triplet state since the substituent groups in molecules in this state are twisted out of plane so as to minimize the overlap between the π and π^* orbitals. The configuration of molecules in this state may be represented approximately by the structure

$$\left[\underset{H}{\overset{R}{>}}\overset{\uparrow}{C}-\overset{\uparrow}{C}\underset{R}{\overset{H}{<}} \right]$$

perpendicular triplet state

The energy of the perpendicular state varies with the angle of twist out of the plane and there will be a certain angle at which the state will have minimum energy. The energy level of the minimum state relative to that of the T_1 state of the *cis* form and that of the *trans* form is depicted schematically in Fig. 5.2.

PHOTOCHEMICAL REACTIONS – I

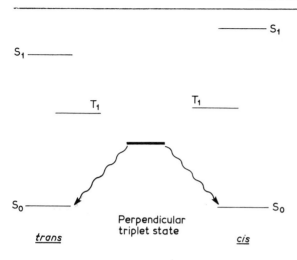

FIGURE 5.2
Relative energies of the S_1 and T_1 states of *cis* and *trans* stilbene and of the 'perpendicular' triplet state of stilbene. The latter state is populated by energy transfer from triplet sensitizers.

Population of the perpendicular triplet state occurs by energy transfer from the triplet state of the sensitizer and since the perpendicular triplet state is of lower energy than the T_1 states of the *cis* and *trans* forms isomerization can be induced using sensitizers with T_1 states of lower energy than the T_1 states of the *cis* and *trans* molecules. This contrasts with systems mentioned in earlier Chapters where energy transfer could only occur efficiently when the energy of the donor triplet state was greater than that of the acceptor triplet state. Deactivation of molecules in the perpendicular triplet state can give either ground state *cis* isomer or ground state *trans* isomer. The formation of these isomers involves rotation of the perpendicular triplet state molecules back to the planar configuration; dependent upon the mode of rotation either the *cis* or *trans* structure will be regained.

The unsensitized isomerizations of simple olefins proceed via the 'normal' triplet states of the *cis* and *trans* forms; these triplet states are populated by intersystem crossing from the corresponding S_1 state. The isomerization step involves rotation around the olefinic bond while the molecule is in the triplet state.

5.5.2 Intramolecular photocyclization

The absorption of radiation by certain olefinic compounds leads to a rearrangment of the bond system and the formation of a cyclized product. Valence bond isomerizations of this type are observed in the following reactions:

(a)

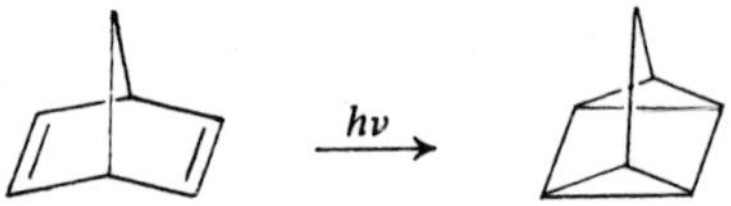

(b)

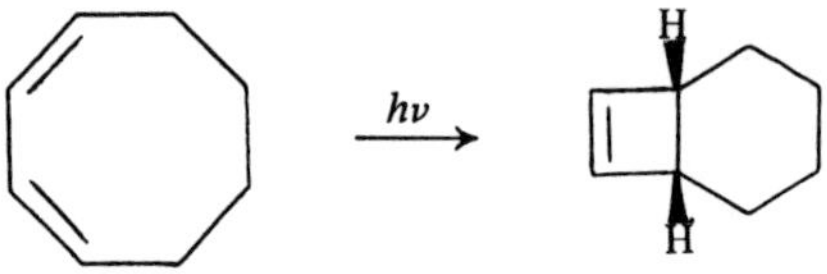

(c)

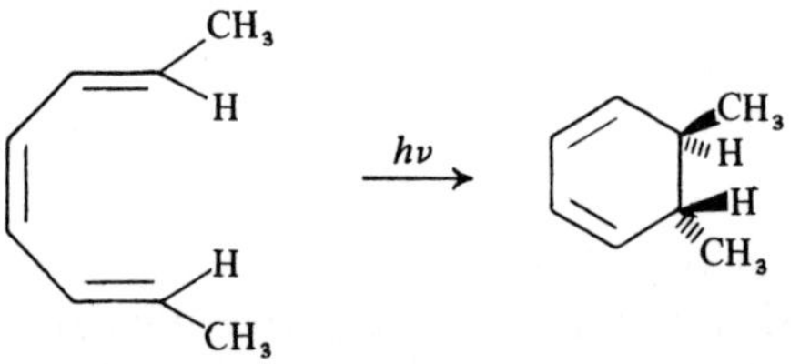

The nature of the products resulting from the photolysis of dienes and trienes is dependent in some cases upon whether reaction takes place from a singlet or triplet state of the photo-excited olefin. For instant, the direct photolysis of 3-methylene-1,5-hexadiene gives a cyclobutene derivative (A) as the product whereas photolysis in the presence of a triplet state sensitizer produces a bicyclo-ring compound (B) (see below). Presumably product (A) is formed from an excited singlet state of the parent olefin while product (B) is formed from a triplet state.

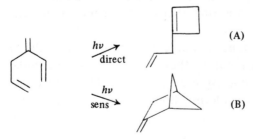

110

PHOTOCHEMICAL REACTIONS – I

Other examples of bond rearrangement accompanied by ring formation are observed in the photolysis of dienones. The photolytic rearrangement of 4,4-diphenylcyclohexa-2,5-diene-1-one has been extensively studied and the following mechanism has been postulated for the reaction:

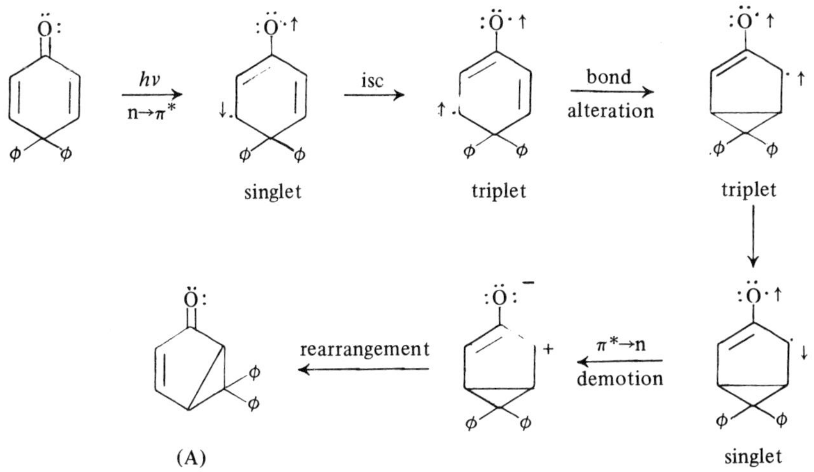

The mechanism invokes the formation of (n, π^*) triplet state molecules which convert to isomeric triplet state molecules by bond alteration. Molecules in the latter triplet state convert by intersystem crossing to singlet state molecules and these are then deactivated to give ground state species. Molecular rearrangement in the ground state molecules then gives the cyclopropyl ketone (A) as the primary photochemical product. This ketone is itself photoactive and is converted to 2,3-diphenylphenol on irradiation.

Acyclic aliphatic ketones can sometimes undergo photocyclization provided there is a hydrogen atom positioned γ to the carbonyl function. The reaction involves intramolecular hydrogen abstraction to give a biradical intermediate which ring closes to form a cyclobutanol derivative:

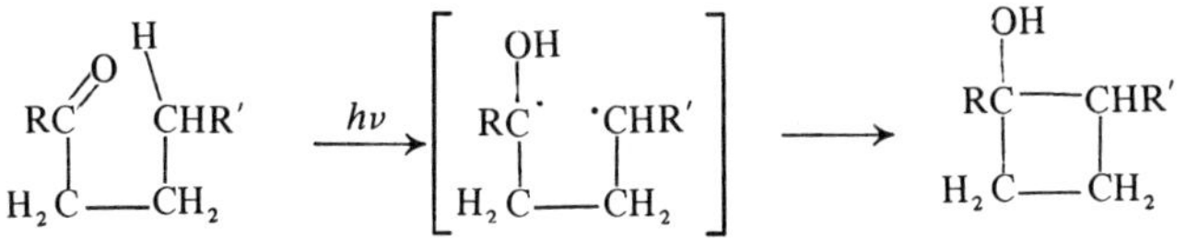

INTRODUCTION TO MOLECULAR PHOTOCHEMISTRY

Reactions of the above type are analogous to the intramolecular hydrogen abstraction reactions of aromatic ketones discussed earlier, and like these reactions proceed via a triplet state of the ketone. In the case of acyclic aliphatic ketones the reactive triplet state will be of the (n, π^*) type.

Cyclizations of the above type can also be induced in 1,2-diketones and keto-ethers:

(a) 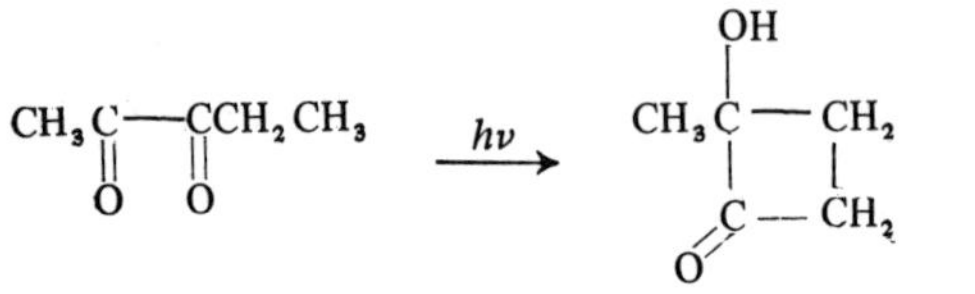 $CH_3C(=O)-C(=O)CH_2CH_3 \xrightarrow{h\nu}$ cyclic product with OH, CH_3, CH_2, CH_2 groups

(b) 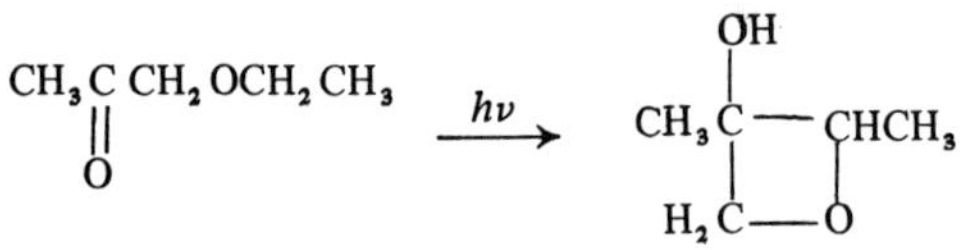 $CH_3C(=O)CH_2OCH_2CH_3 \xrightarrow{h\nu}$ cyclic product with OH, CH_3, $CHCH_3$, H_2C-O groups

5.5.3 Photo-Fries rearrangement

The photo-Fries rearrangement involves migration of a group R across a double bond in structures of the following type:

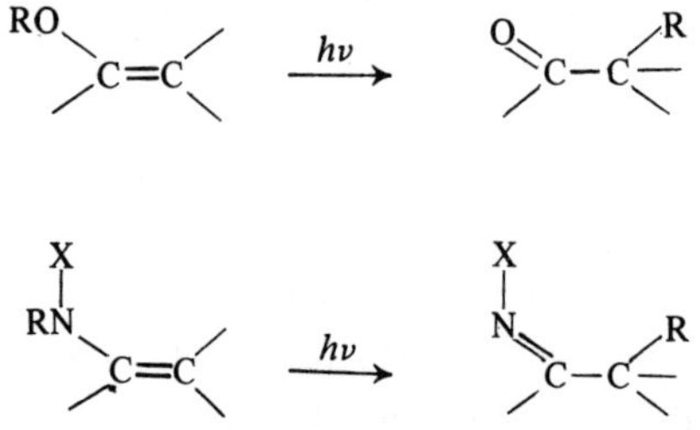

$RO-C=C \xrightarrow{h\nu} O=C-C(R)$

$R_2N(X)-C=C \xrightarrow{h\nu} N(X)=C-C(R)$

When such structures are incorporated in aromatic compounds 1,3 and 1,5 migration of the group R occurs, and this is then followed by a non-photochemical aromatization step. The steps in the reaction of aromatic ethers are shown below:

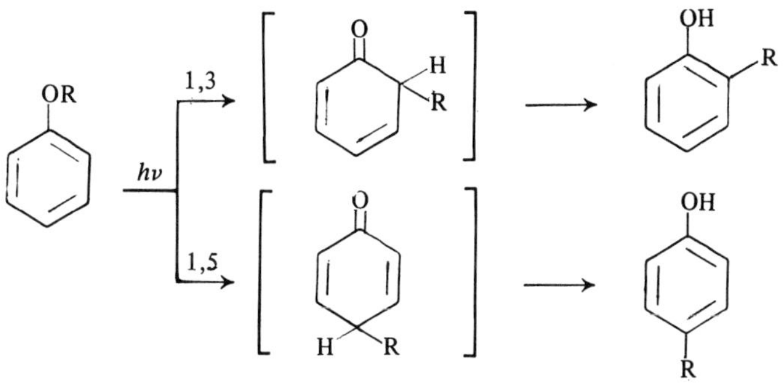

PHOTOCHEMICAL REACTIONS – I

[Reaction scheme: ArOR undergoes hv via 1,3 and 1,5 migration to give ortho- and para-hydroxy R-substituted products]

The corresponding reaction for anilides is as follows:

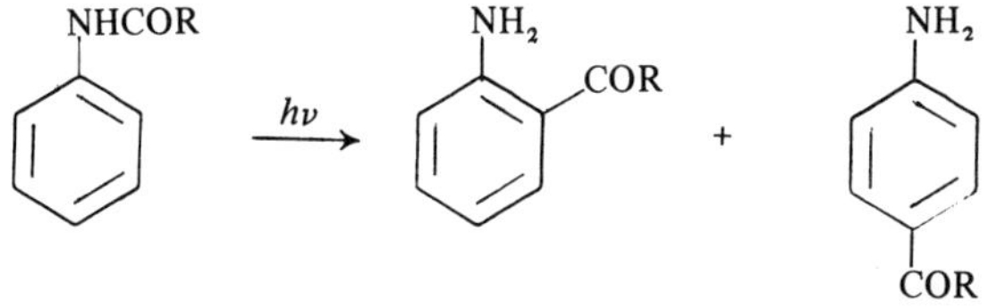

[Reaction scheme: PhNHCOR → hv → ortho-aminoaryl ketone (NH₂, COR) + para-aminoaryl ketone]

For non-aromatic structures only the 1,3 migration step can occur. The double bond involved in the 1,3 migration in such cases can be either isolated or conjugated as shown in the examples below:

(a) 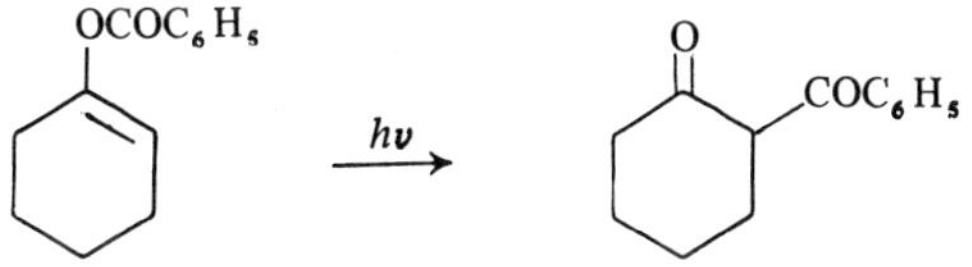 [Cyclohexenyl benzoate (OCOC$_6$H$_5$) → hv → 2-benzoylcyclohexanone (COC$_6$H$_5$)]

(b) 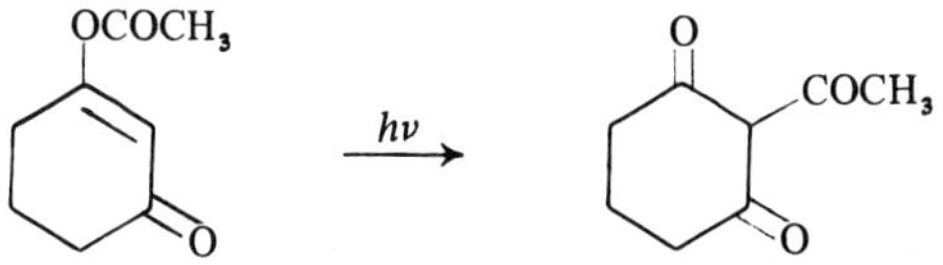 [Cyclohexenone acetate (OCOCH$_3$) → hv → 2-acetylcyclohexane-1,3-dione (COCH$_3$)]

The common feature of these structural isomerization reactions is that they are not affected by either triplet state sensitizers or triplet state quenchers. This suggests that an excited singlet state of the absorbing molecule is involved in the migration step.

5.5.4 Rearrangement of the Benzene Ring

The absorption spectrum of benzene shows a relatively weak $\pi \to \pi^*$ band at 250 nm (ϵ = 20 m² mol⁻¹) and a strong $\pi \to \pi^*$ band at 200 nm (ϵ = 740 m² mol⁻¹). Photolysis of liquid benzene with radiation corresponding to these bands brings about the interesting photochemical transformations shown below:

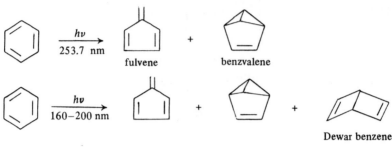

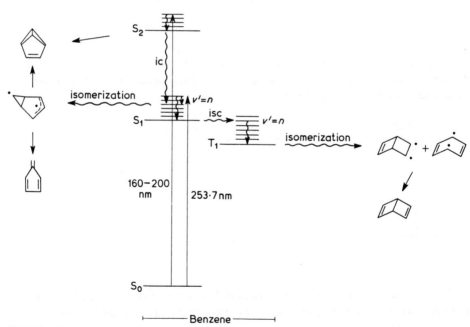

FIGURE 5.3
Mechanism for the conversion of benzene into benzvalene, fulvene and Dewar benzene

PHOTOCHEMICAL REACTIONS — I

The mechanism whereby benzvalene, fulvene and Dewar benzene are formed in the photolysis of benzene is summarized in Fig. 5.3. The biradical intermediates shown in this Figure have been mentioned earlier with regard to photo-addition reactions of benzene (pages 103–104).

Photochemical reactions—II   6

The energy of radiation in the ultraviolet-visible region of the spectrum is comparable with bond dissociation energies in organic molecules and in certain inorganic molecules such as the halogens and hydrogen halides (Fig. 6.1). Consequently it is to be expected that the absorption of ultraviolet-visible radiation could lead to bond fission and to the fragmentation

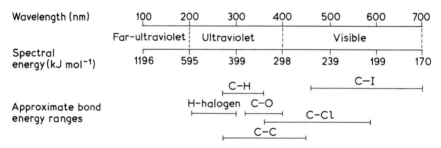

FIGURE 6.1
Comparison of spectral energies in the ultraviolet-visible region with bond dissociation energies

of compounds into lower molecular weight species. Such fragmentation can be of synthetic utility in that radicals may be formed in the primary dissociative step and these may react either with each other or with other species in the system to produce the desired product. Alternatively, the primary dissociative step may involve molecular elimination from the parent compound with the direct formation of the product.

The mechanism of bond dissociation and examples of dissociative and elimination processes in various types of compound and some of their synthetic applications are considered in this Chapter.

6.1 Bond dissociation

The processes whereby the absorption of radiant energy leads to bond breaking in polyatomic molecules can be discussed by reference to the potential energy curves of a diatomic molecule. The potential energy curves for the ground state and an excited state of iodine are represented in Fig. 6.2. The horizontal dotted line across the curve for the excited state represents

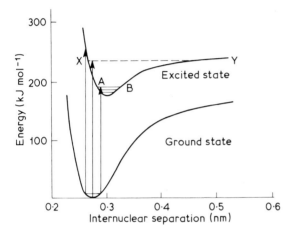

FIGURE 6.2
Potential energy curves for the ground state and an excited state of iodine

the vibrational level in which the excited iodine molecule undergoes a vibration which leads to the rupture of the I–I bond. This vibration can be envisaged by considering the change in internuclear separation between the atoms as the potential energy changes along the curve from the point X to the point Y. At the point Y the internuclear separation can increase indefinitely and the I—I bond will break. This fragmentation process is termed *optical dissociation* since it occurs directly after an optical transition.

Transitions to points on the curve above X will also lead to dissociation since the molecule, after excitation, will undergo a vibration and the I—I bond will break when the internuclear separation reaches the value represented by the point Y. Excitation to the vibrational level AB will not cause bond rupture since the internuclear separation cannot increase beyond the value represented by the point B. Similarly, excitation to any of the vibrational levels below the level XY will be ineffective with regard to bond rupture.

Molecules can have a number of different excited electronic states and as the complexity of the molecule increases so the number of states increases. This means that there will be an increase in the probability of intersection of the potential energy curves of the different states as the molecular complexity increases, and consequently an increased probability of crossing from one excited state to another. Thus it is possible to promote a molecule into a stable vibrational level of an excited state but the molecule may still dissociate because of crossing to another excited state. Dissociation which occurs after a crossing step is referred to as *predissociation*. Possible arrangements of the excited states of a diatomic molecule which can lead to predissociation are shown in Fig. 6.3.

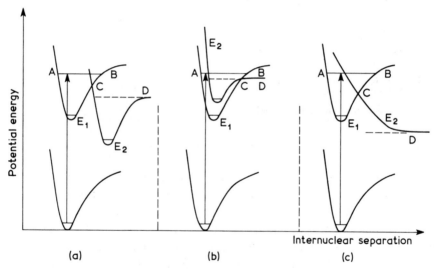

FIGURE 6.3
Arrangements of excited states which lead to predissociation after excitation

Excitation to the vibrational level AB of the excited states E_1 shown in Fig. 6.3 will give a molecule in which the vibration mode will have extremes of internuclear separation given by the points A and B. When the internuclear separation of the vibrating molecule equals the value corresponding to the point C the potential energy of the system is the same as that of the excited state E_2 and cross-over to the state E_2 will occur. In the situation shown in Fig. 6.3(a) the molecule, once in the E_2 state, will undergo a vibration represented by traversing the curve from C to D, and at point D the internuclear bond will break. When the potential energy curves are positioned as in Fig. 6.3(b) the molecule will dissociate immediately on cross-over at point C. If the cross-over point is at lower energy than the dissociative limit of the E_2 state then dissociation will not occur while the molecule is in this state. It may happen that the E_2 state does not have a potential energy minimum at stable internuclear separations (Fig. 6.3(c)) and here cross-over to the E_2 state will lead immediately to dissociation of the molecule.

The photodissociation of polyatomic molecules cannot be represented on a two-dimensional diagram as for the dissociation of diatomic molecules, but nevertheless the principle of the dissociation mechanism is the same. Photolysis of a molecule using radiation of wavelength corresponding to an absorption band of the molecule results in the formation of an excited state. Crossings to a number of other states may occur after excitation and if at any point the relationship between the crossing states is as given in Fig. 6.3(a) to (c) then scission of one of the bonds in the molecule may ensue.

6.2 Photofragmentation and related reactions in various types of compound

The nature of the products formed as a result of photofragmentation is often dependent upon whether the photolysis is performed in the vapour phase or in solution. Radicals resulting from a primary dissociative step $R-R' \rightarrow R^{\cdot} + R^{\cdot\prime}$ are often formed in a high vibrational level of their ground state, and in such cases are referred to as *hot* radicals. In the vapour phase the *hot* radicals are likely to react and form products before a collision can occur with another radical or molecule. When the radicals are formed in solution they are surrounded by a 'solvent cage' and since this 'cage' hinders their separation they may collide with one another before reacting and the original molecule may be reformed. Alternatively, the initially formed *hot* radicals may be deactivated by collision with solvent molecules and the resulting deactivated radicals may react to give products which are not

INTRODUCTION TO MOLECULAR PHOTOCHEMISTRY

formed in the corresponding vapour phase reaction. The liquid phase reaction can also differ from the vapour phase reaction in that the radicals may react chemically with the solvent. For a hydrogen-containing solvent radicals may abstract hydrogen to form products of the type R—H and R′—H.

Examples of vapour and liquid phase photofragmentations in different types of compound are considered in the following sections. Some examples of reactions resulting from photofragmentations are also given.

6.2.1 *Halogens and hydrogen halides*

The halogens — chlorine, bromine, iodine — are coloured compounds and the absorption of visible radiation causes their dissociation into halogen radicals. Generation of these radicals in the presence of an olefin can initiate a chain reaction which results in a di-halogenated product being formed. The mechanism for the reaction is shown below:

Initiation $\qquad X_2 \xrightarrow{h\nu} 2X^\bullet$

Chain reaction $\qquad X^\bullet + R_2C=CR_2 \rightarrow R_2C\overset{\bullet}{X}CR_2$

$\qquad\qquad\qquad\qquad R_2CXCR_2 + X_2 \rightarrow R_2CXCXR_2 + X^\bullet$

The reactivity of halogens in such addition reactions decreases in the order chlorine > bromine > iodine, and usually only chloro and bromo addition products are prepared by this photochemical route. Examples of photohalogenation reactions are listed below:

(a) $\qquad CH_2=CHCl + Cl_2 \xrightarrow{h\nu} CH_2ClCHCl_2$

(b) $\qquad CF_2=CH_2 + Cl_2 \xrightarrow{h\nu} CF_2ClCH_2Cl$

(c) $\qquad CH_2=CHCl + Br_2 \xrightarrow{h\nu} CH_2BrCHBrCl$

(d) $\qquad (CH_3)_3C.C\equiv CH + Br_2 \xrightarrow{h\nu} (CH_3)_3CCBr=CHBr$

Hydrogen halides dissociate on absorption of ultraviolet radiation and here again this can be used as a means of inducing addition to unsaturated systems. Normally, however, this particular addition process is limited in its practical use to hydrogen bromide. The mechanism for the addition to an olefin is as follows:

PHOTOCHEMICAL REACTIONS – II

$$HBr \xrightarrow{h\nu} H^\bullet + Br^\bullet$$

$$Br^\bullet + RCH=CH_2 \longrightarrow R\dot{C}HCH_2Br$$

$$R\dot{C}HCH_2Br + HBr \longrightarrow RCH_2CH_2Br + Br^\bullet$$

The addition of hydrogen bromide to olefins and acetylenes is a useful route for the preparation of alkyl and alkenyl bromides:

(a) $\qquad CH_3CH=CH_2 + HBr \xrightarrow{h\nu} CH_3CH_2CH_2Br$

(b) $\qquad CH_2=CCl_2 + HBr \xrightarrow{h\nu} CH_2BrCHCl_2$

(c) $\qquad CH\equiv CH + HBr \xrightarrow{h\nu} CH_2=CHBr$

(d) $\qquad CH_3C\equiv CH + HBr \xrightarrow{h\nu} CH_3CH=CHBr$

6.2.2 Organic halides

Photolysis of alkyl halides using radiation of a wavelength corresponding to their first absorption band (200–300 nm) causes the C-halogen bond to break:

$$RX \xrightarrow{h\nu} R^\bullet + X^\bullet$$

In the case of the liquid phase photolysis of alkyl iodides the final products are molecular iodine and an equimolar mixture of the corresponding alkane and alkene. These products are explicable in terms of the following reactions:

$$RCH_2CH_2I \xrightarrow{h\nu} RCH_2CH_2^\bullet + I^\bullet$$

$$RCH_2CH_2^\bullet + RCH_2CH_2I \longrightarrow RCH_2CH_3 + R\dot{C}HCH_2I$$

$$R\dot{C}HCH_2I \longrightarrow RCH=CH_2 + I^\bullet$$

$$I^\bullet + I^\bullet \longrightarrow I_2$$

Photolysis of alkyl halides with radiation corresponding to their second absorption band (170–200 nm) may result in the elimination of hydrogen halide from the parent compound:

$$\text{RCH}_2\text{CH}_2\text{X} \xrightarrow{h\nu} \text{RCH}=\text{CH}_2 + \text{HX}$$

Irradiation of aromatic halides induces the cleavage of the C-halogen bond in the primary step with the formation of an aryl radical and a halogen atom:

$$\text{C}_6\text{H}_5\text{X} \xrightarrow{h\nu} \text{C}_6\text{H}_5^{\cdot} + \text{X}^{\cdot}$$

The photolysis of *ortho*-diiodobenzene is of particular interest in that both iodine atoms are eliminated and benzyne is formed:

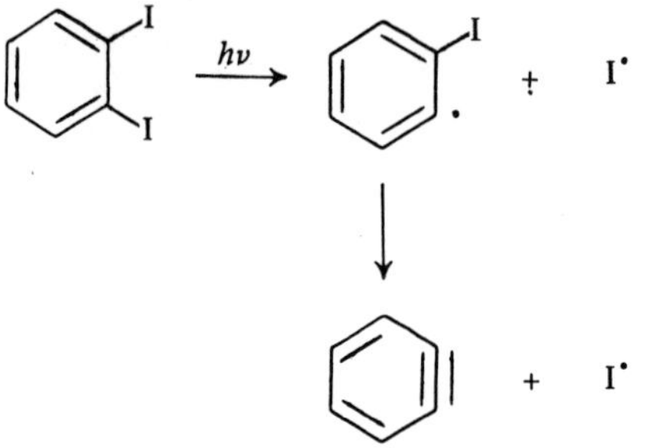

Benzyne is unstable but it can be trapped out from the reaction mixtures as a component of a Diels–Alder adduct.

Halogenated aliphatic compounds can be synthesized by photolyzing polyhalomethanes in the presence of an olefin. The addition of carbon tetrabromide and carbon tetrachloride to terminal olefins typifies the reaction:

(a) $\quad \text{RCH}=\text{CH}_2 + \text{CBr}_4 \xrightarrow{h\nu} \text{RCHBrCH}_2\text{CBr}_3$

(b) $\quad \text{RCH}=\text{CH}_2 + \text{CCl}_4 \xrightarrow{h\nu} \text{RCHClCH}_2\text{CCl}_3$

Irradiation of mixed polyhalomethanes such as bromotrichloromethane and trifluoroiodomethane leads to preferential rupture at the weakest C-halogen bond, and again in the presence of olefins 1 : 1 adducts are formed:

(a) $\quad \text{BrCCl}_3 \xrightarrow{h\nu} \text{Br}^{\cdot} + {}^{\cdot}\text{CCl}_3 \xrightarrow{\text{RCH}=\text{CH}_2} \text{RCHBrCH}_2\text{CCl}_3$

(b) $\quad \text{ICF}_3 \xrightarrow{h\nu} \text{I}^{\cdot} + {}^{\cdot}\text{CF}_3 \xrightarrow{\text{RCH}=\text{CH}_2} \text{RCHICH}_2\text{CF}_3$

6.2.3 Hydrocarbons

The lowest energy absorption in paraffin hydrocarbons occurs in the vacuum ultraviolet region (120–200 nm) and photolysis with radiation within this band causes the breakdown of the hydrocarbon. The main fragmentation process results in the elimination of hydrogen:

$$RCH_2R \xrightarrow{h\nu} R\ddot{C}R + H_2$$

Photolysis of olefinic hydrocarbons in the vapour phase gives rise to a number of reactions with hydrogen elimination as one of the major processes. In contrast, the photolysis of olefins in the liquid phase often only results in isomerization:

Vapour phase:
$$\underset{R}{\overset{H}{\diagdown}}C=C\underset{H}{\overset{H}{\diagup}} + h\nu \rightarrow RC\equiv CH + H_2$$
$$\rightarrow RCH=C\colon + H_2$$
$$\rightarrow R\dot{C}=CH_2 + H^\bullet$$
$$\rightarrow RCH=\dot{C}H + H^\bullet$$
$$\rightarrow {}^\bullet CH=CH_2 + R^\bullet$$

Solution:
$$\underset{R}{\overset{H}{\diagdown}}C=C\underset{H}{\overset{R}{\diagup}} + h\nu \rightarrow \underset{H}{\overset{R}{\diagdown}}C=C\underset{H}{\overset{R}{\diagup}}$$

Aromatic hydrocarbons are more photostable than aliphatic hydrocarbons and irradiation within the longest wavelength band gives very low yields of photodecomposition products. For instance, irradiation of benzene in the vapour phase with 253.7 nm radiation (first band) does not cause any significant decomposition, whereas using radiation in the region 185–200 nm (second and third bands) causes considerable decomposition and the production of acetylene, hydrogen, methane, ethane and polymeric materials.

The photolysis of alkyl benzenes with radiation of wavelength corresponding to the first absorption band causes bond rupture in the alkyl group. The primary reactions taking place during the photolysis of ethylbenzene with 253.7 nm radiation are as follows:

$$C_6H_5CH_2CH_3 + h\nu \rightarrow C_6H_5\dot{C}HCH_3 + H^\bullet$$
$$\rightarrow C_6H_5CH=CH_2 + H_2$$
$$\rightarrow C_6H_5\dot{C}H_2 + \dot{C}H_3$$

Hydrocarbons containing cyclobutane ring systems often dissociate into smaller fragments molecules on absorbing radiant energy. For example, the bicyclo [4,2,0] octa-2,4-diene system (A) has been found to fragment to give either benzene and an olefin or an acyclic tetraene, depending upon the reaction conditions used:

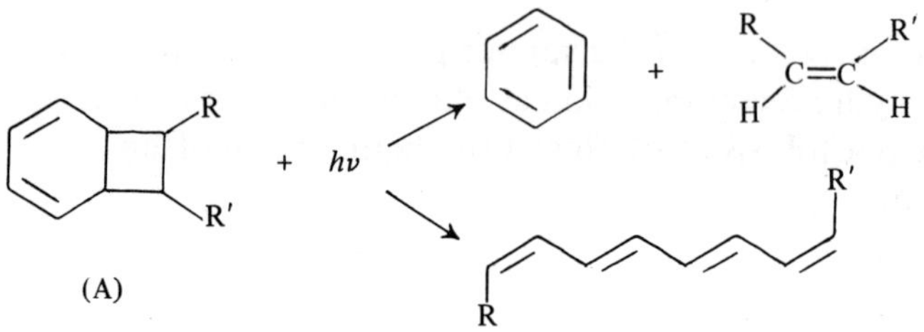

6.2.4 Aldehydes and ketones

Photolysis of aliphatic carbonyl compounds with radiation corresponding to their longest wavelength absorption band (270–300 nm) gives rise to products formed via bond fragmentation. There are two predominant processes whereby these products are formed viz. the Norrish Type I process and the Norrish Type II process.

The *Norrish Type I* process involves cleavage of a bond α to the carbonyl group, followed by elimination of CO from the resulting acyl radical. The two possibilities for the process in ketones of the type RCOR′ are as follows:

(a) $\qquad$ RCOR′ $\xrightarrow{h\nu}$ RCO$^{\bullet}$ + R′$^{\bullet \bullet}$ → R$^{\bullet}$ + R″$^{\bullet \bullet}$ + CO

(b) $\qquad$ RCOR′ $\xrightarrow{h\nu}$ R$^{\bullet}$ + R′CO$^{\bullet}$ → R$^{\bullet}$ + R″$^{\bullet \bullet}$ + CO

Rupture of the bond α to the carbonyl group in cyclic ketones produces a biradical which after elimination of CO can either fragment or cyclize. This is seen in the mechanism given below for the vapour phase photolysis of cyclopentanone:

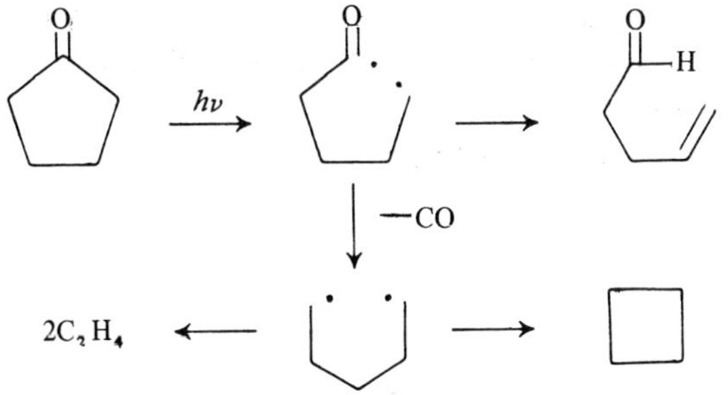

The elimination of CO from carbonyl compounds occurs readily in the vapour phase but does not occur in the liquid phase except in the photolysis of small or medium ring compounds. That is, liquid phase decarbonylation is dependent upon there being some degree of strain in the ring from which the CO molecule is eliminated. The cyclobutane ring in 2,2,4,4-tetramethyl-cyclobutane-1,3-dione is severely strained and photolysis of this compound in benzene solution leads to a step-wise loss of two molecules of carbon monoxide:

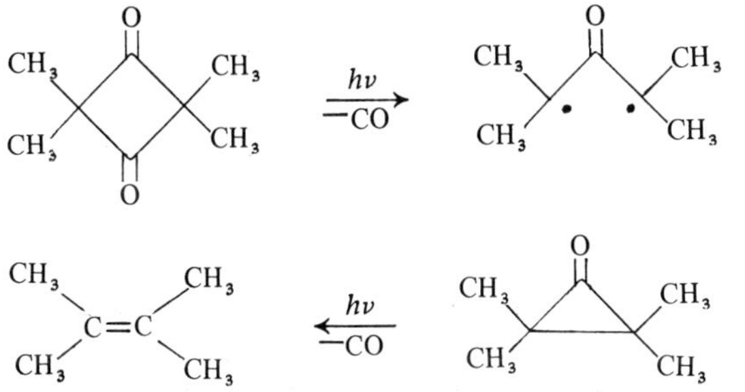

Evidence for the participation of the biradical (A) in the above reaction sequence was gained by trapping it out as a component of an adduct with furan:

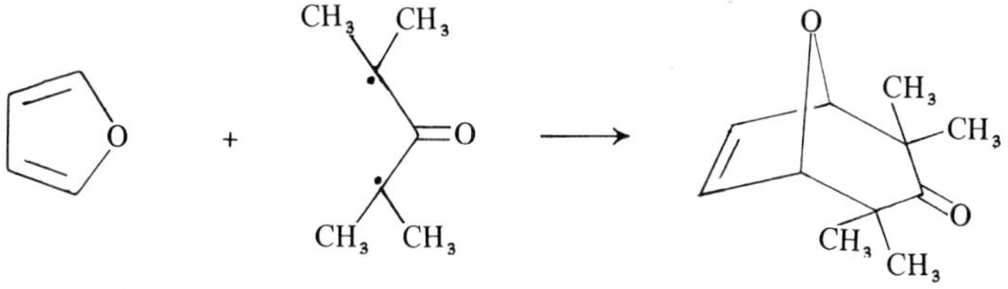

Irradiation of strained cis-α-diketones can also cause decarbonylation and again the observed products can be rationalized by a mechanistic scheme involving initial rupture of the α carbon-carbon bond. For example, irradiation of 3,4-di-t-butylcyclobutanedione yields di-t-butylcyclopropanone:

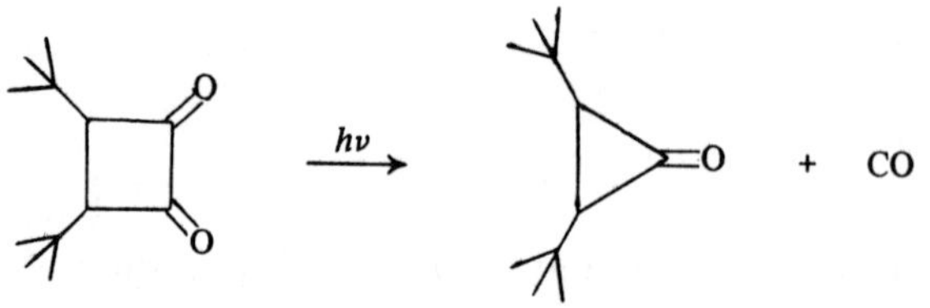

Other examples of liquid phase decarbonylation reactions are seen below:

(a)

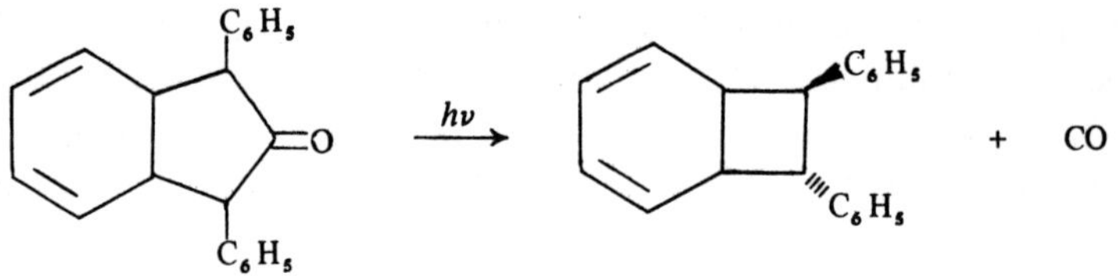

(b)

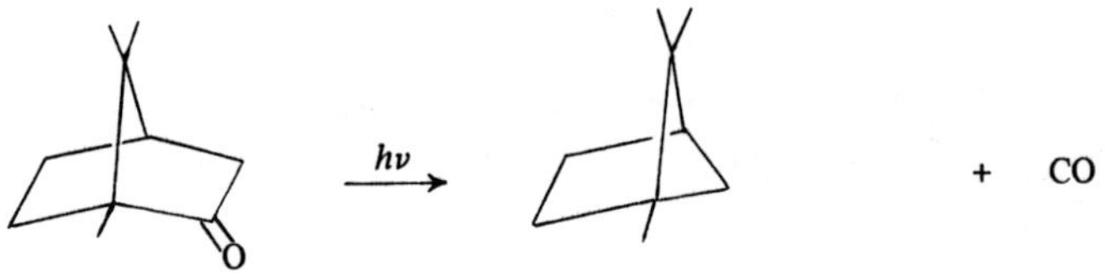

The *Norrish Type II* process involves the formation of a six-membered transition state in which a γ-hydrogen atom is 'bonded' to the oxygen atom of the carbonyl group. Bond breaking in the transition state leads to the formation of the reaction products. In the case of acyclic aliphatic compounds the products are olefins and methyl carbonyl compounds:

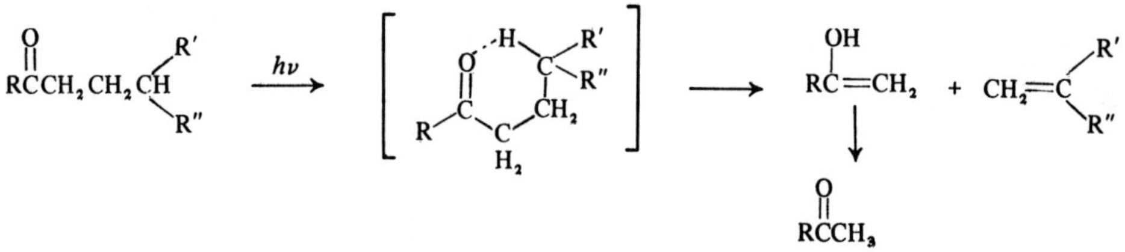

126

Ketones possessing a γ-hydrogen atom may react additionally via intramolecular hydrogen abstraction to give a cyclobutanol derivative:

$$\left[\begin{array}{c}\text{H} \quad \text{R}' \\ \text{O} \quad \text{C} \\ \| \quad | \\ \text{C} \quad \text{C}-\text{R}'' \\ \text{R} \quad \text{C} \quad \text{CH}_2 \\ \text{H}_2\end{array}\right] \longrightarrow \left[\begin{array}{c}\text{OH} \quad \text{C}-\text{R}' \\ | \quad | \\ \text{C}\cdot \quad \text{C}-\text{R}'' \\ \text{R} \quad \text{C} \quad \text{CH}_2 \\ \text{H}_2\end{array}\right] \longrightarrow \begin{array}{c}\text{HO} \quad \text{R}' \\ \text{C}---\text{C} \\ \text{R} \quad | \quad | \quad \text{R}'' \\ \text{H}_2\text{C}---\text{CH}_2\end{array}$$

Although numerous investigations have been carried out on Norrish Type I and Type II reactions in most cases it is not yet known whether the photo-excited ketone reacts while in the S_1 state or in the T_1 state. There is some evidence that for compounds undergoing the Type II process the T_1 state of the ketone is involved. For instance, it has been found that methylated derivatives of acetophenone which have T_1 states of the (n, π^*) type undergo the Type II process more readily than derivatives which have T_1 states of the (π, π^*) type.

6.2.5 Acids and esters

Irradiation of aliphatic acids and esters within their longest wavelength absorption band causes extensive bond rupture within the functional group. The sites at which scission occurs are shown below:

$$R \overset{1}{\dashv} \overset{\text{O}}{\underset{\|}{C}} \overset{2}{\dashv} O \overset{3}{\dashv} H \qquad R \overset{1}{\dashv} \overset{\text{O}}{\underset{\|}{C}} \overset{2}{\dashv} O \overset{3}{\dashv} R$$

The fragmentation of acids and esters leads to the formation of radicals which subsequently react to give hydrocarbons, olefins, carbon dioxide and carbon monoxide as products.

If an ester, RCOOR', has a γ-hydrogen in the acid section (R) or a β-hydrogen in the alcohol section (R') then photochemical reaction may take place by either of the Norrish Type II processes (a) or (b) illustrated below.

(a) $\text{R'OCCH}_2\text{CH}_2\text{CH}_3 \xrightarrow{h\nu} \left[\begin{array}{c}\text{H} \\ \text{O} \quad \text{CH}_2 \\ \| \quad | \\ \text{C} \quad \text{CH}_2 \\ \text{R'O} \quad \text{C} \\ \text{H}_2\end{array}\right] \longrightarrow \text{R'O C=CH}_2 + \text{CH}_2\text{=CH}_2$

$$\downarrow$$

$$\underset{\text{R'OCCH}_3}{\overset{\text{O}}{\|}}$$

INTRODUCTION TO MOLECULAR PHOTOCHEMISTRY

(b) 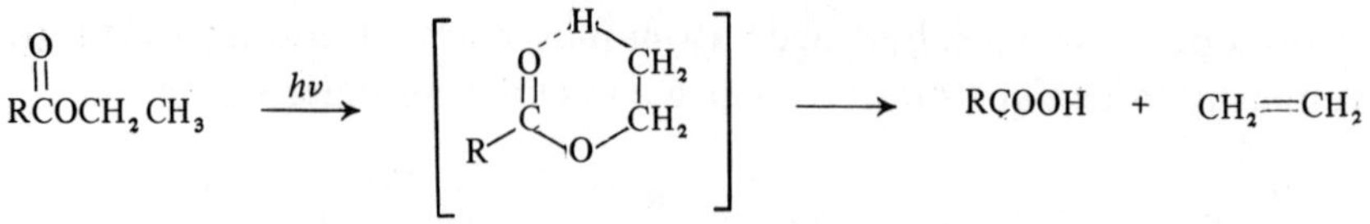 $RCOCH_2CH_3 \xrightarrow{h\nu} \left[\begin{array}{c} \text{cyclic transition state} \end{array} \right] \longrightarrow RCOOH + CH_2{=}CH_2$

In certain instances the elimination of carbon dioxide from an acid may take place by a concerted reaction rather than by a reaction involving radicals. For example, the mechanism for the photodecomposition of pyruvic acid in the vapour phase has been postulated as

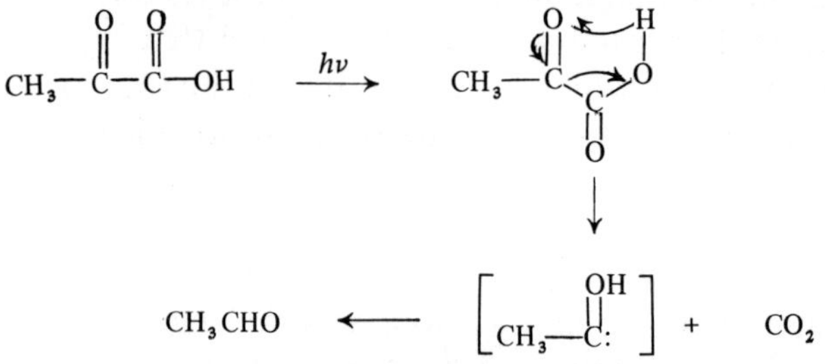

$$CH_3-\underset{\parallel}{C}-\underset{\parallel}{C}-OH \xrightarrow{h\nu} CH_3-C\cdots$$

$$\downarrow$$

$$CH_3CHO \longleftarrow \left[CH_3-\underset{\parallel}{C}:\overset{OH}{} \right] + CO_2$$

Photodecarboxylation can also take place in the liquid phase as instanced by the reactions of 1-naphthyl acetic acid and n-butyric acid:

(a) 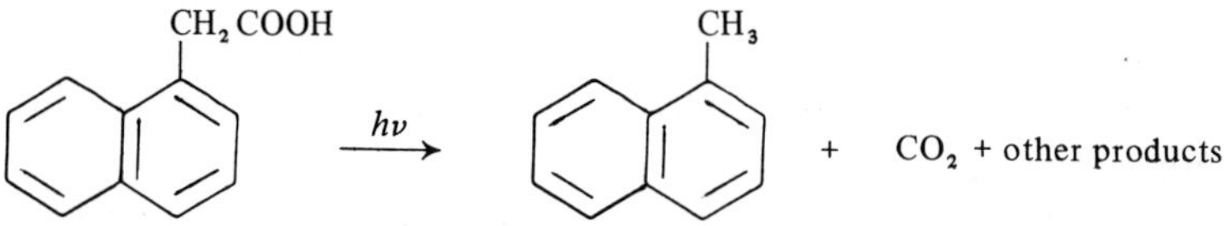 naphthyl-$CH_2COOH \xrightarrow{h\nu}$ naphthyl-CH_3 + CO_2 + other products

(b) $CH_3CH_2CH_2COOH \xrightarrow{h\nu} CH_3CH_2CH_3 + CO_2$

6.2.6 Alcohols and ethers

The vapour phase photolysis of aliphatic alcohols and ethers with short wavelength radiation (< 200 nm) causes rupture of the C—O and O—H bonds and the formation of radicals and molecular elimination products.

PHOTOCHEMICAL REACTIONS – II

This is seen below for the photofragmentation of n-propanol:

$$CH_3CH_2CH_2OH + h\nu \rightarrow CH_3CH_2CH_2^\bullet + OH^\bullet$$
$$\rightarrow CH_3CH_2CH_2O^\bullet + H^\bullet$$
$$\rightarrow CH_3CH_2CHO + H_2$$
$$\rightarrow CH_3CH_2CH_3 + H_2O$$
$$\rightarrow CH_3CH_3 + CH_2O$$

The liquid phase photolysis of aliphatic alcohols and ethers does not lead normally to such a complexity of reactions as above. Alcohols and ethers possess a potentially reactive C—H bond α to the oxygen atom and elimination of a hydrogen atom from this site either by direct photolysis or by abstraction by a radical or a photo-activated ketone is a common occurrence in liquid phase reactions. The free radical generated in this step may then initiate a chain reaction which results in the addition of the alcohol or ether to the olefin. This is exemplified below for the photo-sensitized addition of tetrahydrofuran to maleic anhydride:

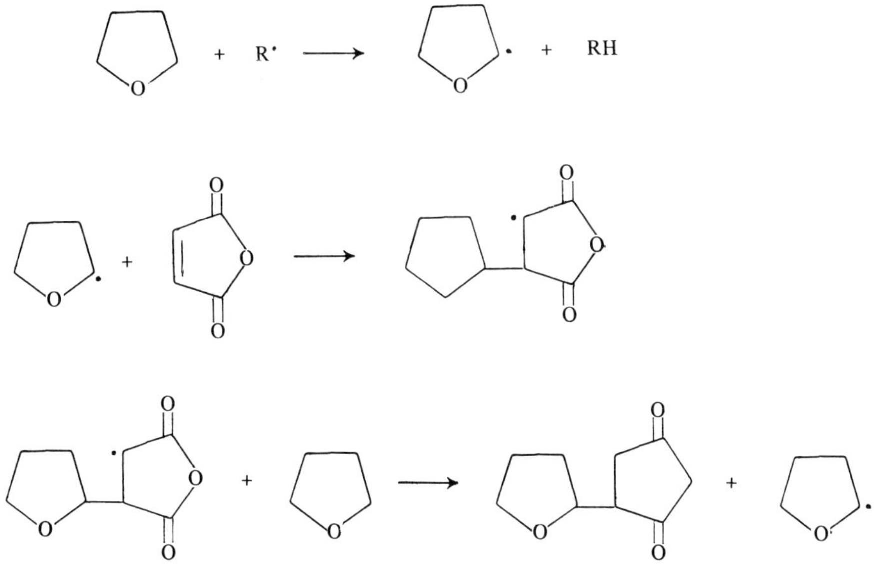

INTRODUCTION TO MOLECULAR PHOTOCHEMISTRY

Other examples of this type of reaction are as follows:

(a) 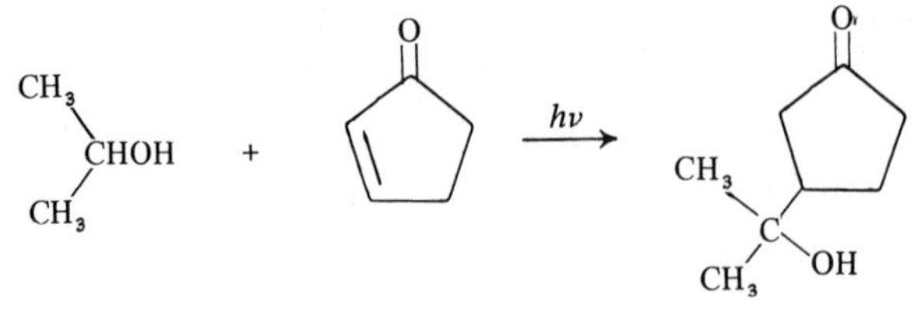 $(CH_3)_2CHOH$ + [cyclopentenone] $\xrightarrow{h\nu}$ [cyclopentanone with $C(CH_3)_2OH$ substituent]

(b) 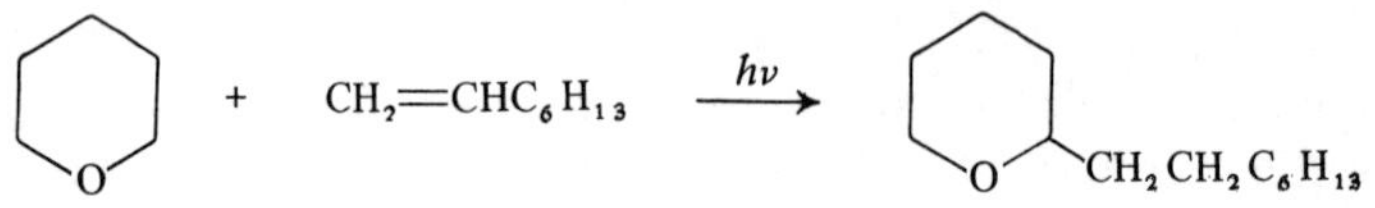 [tetrahydropyran] + $CH_2{=}CHC_6H_{13}$ $\xrightarrow{h\nu}$ [tetrahydropyran with $CH_2CH_2C_6H_{13}$ substituent]

6.2.7 Peroxides

Alkyl peroxides rupture at the weak RO—OR′ bond on absorbing radiation of wavelengths corresponding to their first absorption band (200–350 nm):

$$RO{-}OR' \xrightarrow{h\nu} R\dot{O}^* + {}^\bullet OR'^*$$

The alkoxy radicals formed in the above step have a large excess of energy and will fragment rapidly unless involved in chemical reaction or stabilized by collision with another species.

When radiation of wavelength less than 230 nm is used for the photolysis of peroxides a second dissociative step becomes important:

$$ROOR' \xrightarrow{h\nu} RO_2^{\bullet *} + R'^{\bullet \bullet}$$

The primary steps in the photofragmentation of di-*t*-butyl peroxide are shown below, and here radical fragmentation in the vapour phase results in acetone and methyl radicals being produced:

$$(CH_3)_3COOC(CH_3)_3 \xrightarrow{h\nu} 2(CH_3)_3C\dot{O}^* \rightarrow CH_3COCH_3 + CH_3^{\bullet}$$

When the photolysis of di-*t*-butyl peroxide is performed in solution the initially formed radicals are deactivated by collision with solvent molecules and the second decomposition step becomes unimportant.

The photolysis of alkyl peroxides is a convenient way of generating alkoxy radicals either for studying their reactions or for initiating other reactions.

PHOTOCHEMICAL REACTIONS – II

One of the most important applications of peroxide photolysis is in the initiation of free-radical addition polymerization. Dibenzyl peroxide is frequently used as a radical source in this respect:

$$(C_6H_5COO)_2 \xrightarrow{h\nu} 2\,C_6H_5CO\dot{O}^* \rightarrow 2\,C_6H_5^\cdot + 2\,CO_2$$

Both the benzoyloxy and phenyl radicals produced in the above reactions can act as polymerization initiators. These radicals (R˙) act by adding to the olefinic bond of a suitable monomer to give a second radical which grows further by adding to another monomer molecule. Radical growth is repeated many times and a polymer chain is built up:

Initiation $\quad\quad R^\cdot + CH_2{=}CHX \rightarrow {}^\cdot CH_2CHXR$

Chain propagation $\quad CH_2{=}CHX + {}^\cdot CH_2CHXR \rightarrow {}^\cdot CH_2CHXCH_2CHXR$

Termination $\quad {}^\cdot CH_2CHX(CH_2CHX)_m CH_2CHXR +$
$\quad\quad\quad\quad\quad {}^\cdot CH_2CHX(CH_2CHX)_n CH_2CHXR \rightarrow$ Polymer

6.2.8 Nitrites

The major photofragmentation process which occurs in alkyl nitrites is the dissociation of the RO—NO bond. This gives rise to the formation of alkoxy radicals and nitric oxide:

$$RO{-}NO \xrightarrow{h\nu} RO^\cdot + NO^\cdot$$

Further reactions of the alkoxy radicals lead to products, e.g. the alkoxy radical from *t*-pentyl nitrite reacts to give acetone and methyl ethyl ketone:

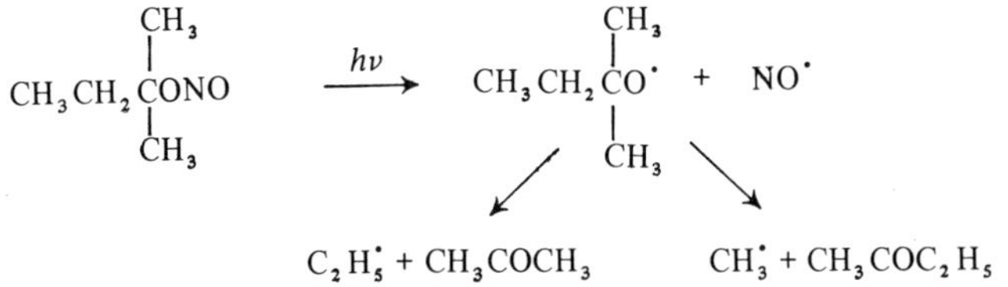

$$CH_3CH_2\underset{\underset{CH_3}{|}}{\overset{\overset{CH_3}{|}}{C}}ONO \xrightarrow{h\nu} CH_3CH_2\underset{\underset{CH_3}{|}}{\overset{\overset{CH_3}{|}}{C}}O^\cdot + NO^\cdot$$

$$C_2H_5^\cdot + CH_3COCH_3 \quad\quad CH_3^\cdot + CH_3COC_2H_5$$

When there is a hydrogen atom in the position δ to the nitrite group then intramolecular hydrogen abstraction may occur within the alkoxy radical. This results in a free radical site at the δ position and if the NO species

INTRODUCTION TO MOLECULAR PHOTOCHEMISTRY

originally eliminated is in the vicinity then radical combination may take place to give an oxime. This sequence is known as the *Barton* reaction. The steps in the reaction are summarized as follows:

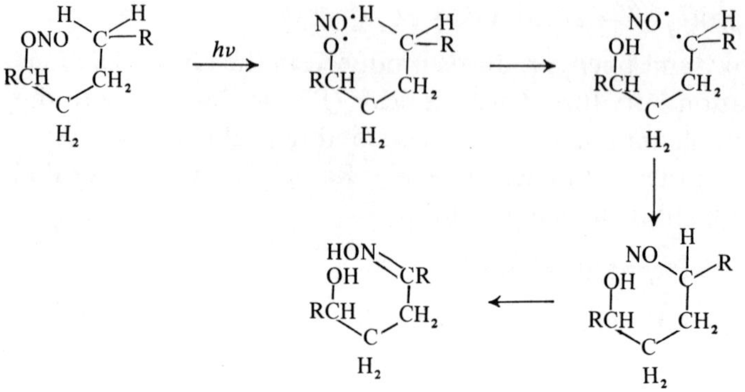

[Reaction scheme showing the Barton reaction mechanism: starting nitrite ester undergoes hv to form excited state, then rearranges through NO• and alcohol intermediates to give the oxime product HON=CR]

The value of the Barton reaction is that it provides a way of activating a saturated carbon centre, and this has proved of great value in steroid synthesis. The following example shows the way in which the reaction can be used:

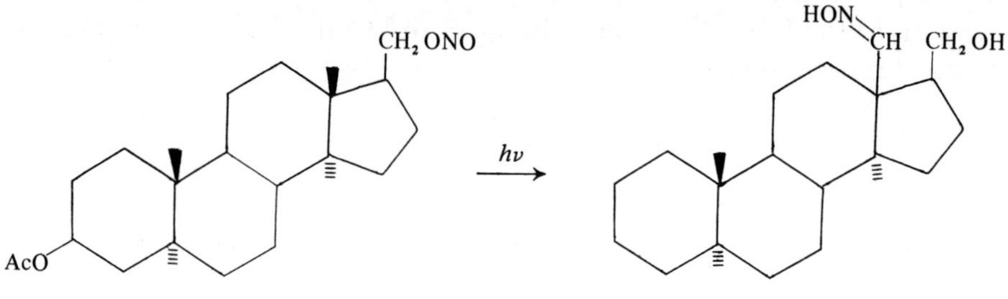

[Steroid structure with CH$_2$ONO and AcO groups undergoes hv to give steroid with HON=CH and CH$_2$OH groups]

6.2.9 Amides

Photolysis of aliphatic amides with radiation in the wavelength region 200–230 nm results in cleavage of either the C—N bond or the C—C bond adjacent to the absorbing carbonyl group:

$$RCONH_2 + h\nu \longrightarrow RCO^\bullet + {}^\bullet NH_2$$
$$\searrow R^\bullet + {}^\bullet CONH_2$$

132

PHOTOCHEMICAL REACTIONS – II

The photofragmentation of amides can be used as a means of achieving their addition to olefins and consequently for synthesizing higher homologous amides. The photoaddition of formamide to olefins has been the most extensively studied of these reactions and it has been found that the addition can be induced either by direct photolysis or by photosensitization using ketones such as acetone or benzophenone. The reaction which is known as 'photoamidation' proceeds via the following free-radical mechanism:

$$HCONH_2 \xrightarrow{h\nu} {}^{\cdot}CONH_2 + H^{\cdot}$$

$$RCH=CH_2 + {}^{\cdot}CONH_2 \longrightarrow R\dot{C}HCH_2CONH_2$$

$$R\dot{C}HCH_2CONH_2 + HCONH_2 \longrightarrow RCH_2CH_2CONH_2 + {}^{\cdot}CONH_2$$

Photoamidation can be applied to a variety of olefin types (terminal, non-terminal, acyclic) and to α,β-unsaturated esters and to acetylenes:

(a) 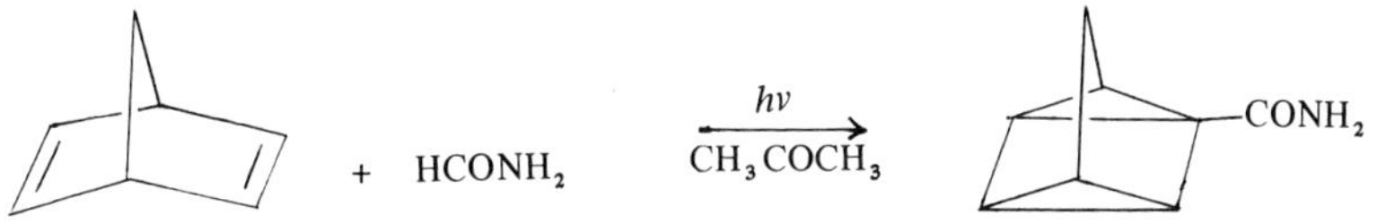 [norbornene structure] + $HCONH_2$ $\xrightarrow[CH_3COCH_3]{h\nu}$ [norbornane–$CONH_2$ structure]

(b) $C_5H_{11}CH=CHCOOR + HCONH_2 \xrightarrow[(C_6H_5)_2CO]{h\nu} C_5H_{11}CHCH_2COOR$
$\qquad\qquad\qquad\qquad\qquad\qquad\qquad\qquad\qquad\qquad\quad |$
$\qquad\qquad\qquad\qquad\qquad\qquad\qquad\qquad\qquad\quad CONH_2$

(c) $C_5H_{11}C{\equiv}CH + HCONH_2 \xrightarrow[(C_6H_5)_2CO]{h\nu}$ $C_5H_{11}\underset{\underset{CONH_2}{|}}{\overset{\overset{CONH_2}{|}}{CH}}CH=CHC_5H_{11}$

6.2.10 Amines

The main photofragmentation processes in primary aliphatic amines are cleavage of the N—H bond with the formation of an amino radical, and cleavage of a C—H bond α to the functional group with the formation of an amine radical:

$$RCH_2NH_2 + h\nu \begin{cases} \longrightarrow RCH_2\dot{N}H + H^{\cdot} \\ \longrightarrow R\dot{C}HNH_2 + H^{\cdot} \end{cases}$$

133

The second of the above processes is of importance in liquid phase photolytic reactions and in the presence of olefins results in 1 : 1 amine-olefin adducts being formed, e.g.

$$CH_3(CH_2)_5NH_2 + CH_2{=}CH(CH_2)_5CH_3 \xrightarrow{h\nu} CH_3(CH_2)_4\overset{\overset{\displaystyle NH_2}{|}}{C}H(CH_2)_7CH_3$$

Aromatic amines are more photostable than aliphatic amines since in aromatic compounds the absorbed radiant energy can be dissipated by radiative emission processes. However photodissociation of aromatic amines does occur to a certain extent especially at elevated temperatures. The major dissociative processes for aniline are shown below:

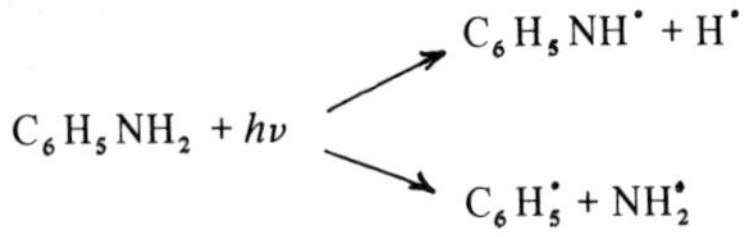

$$C_6H_5NH_2 + h\nu \begin{cases} \longrightarrow C_6H_5NH^\bullet + H^\bullet \\ \longrightarrow C_6H_5^\bullet + NH_2^\bullet \end{cases}$$

6.2.11 Azo-compounds

Azo-alkanes when photolyzed in the vapour phase are readily fragmented into alkyl radicals and nitrogen but in the liquid phase molecular fragmentation is a minor reaction and isomerization takes over as the main photolytic process, e.g.

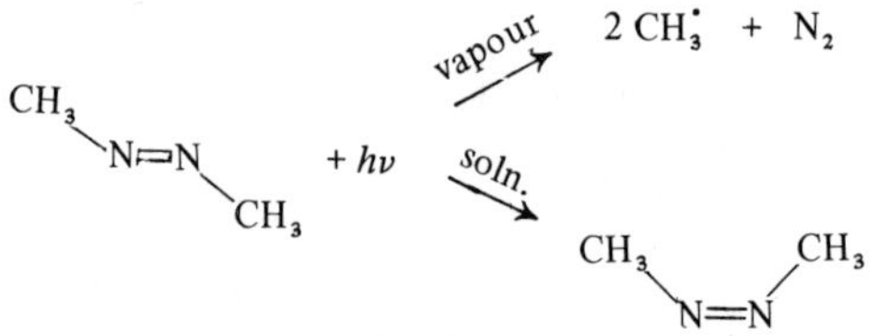

$$\underset{CH_3}{\overset{CH_3}{\diagdown}}N{=}N + h\nu \begin{cases} \xrightarrow{\text{vapour}} 2\,CH_3^\bullet + N_2 \\ \xrightarrow{\text{soln.}} \underset{N{=}N}{\overset{CH_3\diagup\diagdown CH_3}{}} \end{cases}$$

Quantitative studies on the photolysis of *cis* and *trans* azoisopropane in the vapour and liquid phases have indicated that for both compounds fragmentation occurs from high vibrational levels of a (π, π^*) triplet state and isomerization from low vibrational levels of a (π, π^*) triplet state.

The photofragmentation of azo-alkanes is a particularly 'clean' reaction in that only alkyl radicals and nitrogen are formed. Since nitrogen is a rather

PHOTOCHEMICAL REACTIONS – II

inert species this is a convenient method for producing alkyl radicals either for studying their reactions or for initiating other reactions such as free-radical polymerization. Azobisisobutyronitrile is the most commonly used radical source:

$$(CH_3)_2 \underset{CN}{C} - N=N - \underset{CN}{C}(CH_3)_2 \xrightarrow{h\nu} 2(CH_3)_2 \underset{CN}{C^{\cdot}} + N_2$$

Cyclic azo-compounds often eliminate nitrogen on irradiation and here the associated bond rupture produces biradicals which can ring close to give cyclized products. A number of such reactions are known, e.g.

(a), (b), 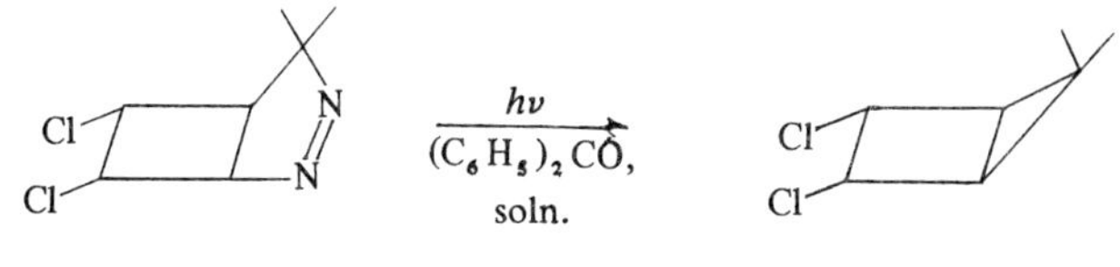 (c) [reaction schemes with structures]

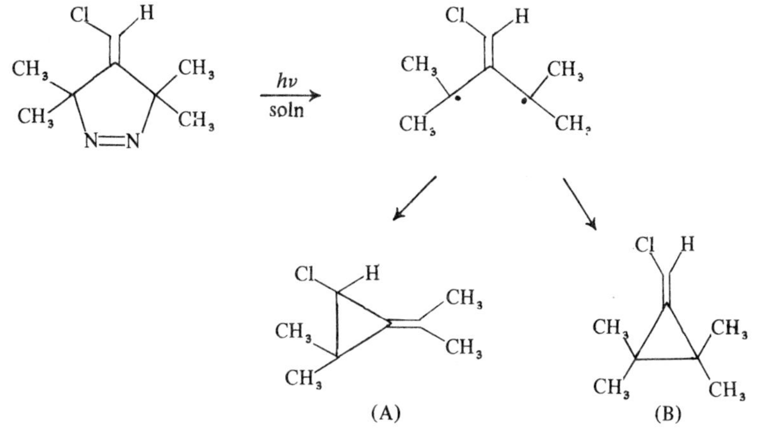

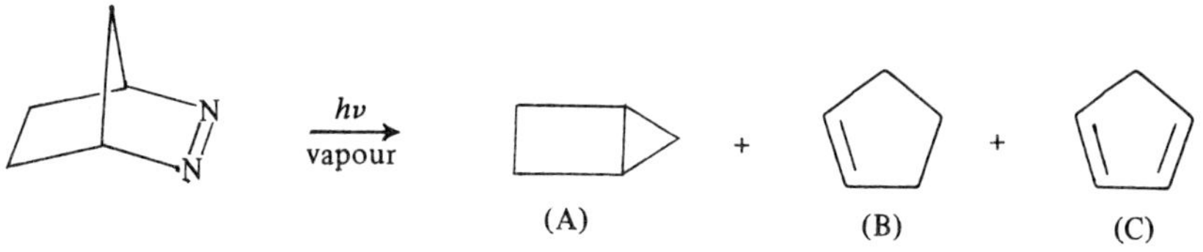

In reaction (a) above, product (A) predominates when the photolysis is performed in the presence of a triplet state sensitizer while product (B) predominates when the azo-compound is photolyzed directly. Thus it is likely that (A) is formed via a triplet biradical species and (B) via a singlet biradical species. Reaction (b) above apparently proceeds via a triplet biradical as evidenced by the fact that the reaction proceeds quantitatively when benzophenone is used as sensitizer. Reaction (c) is of interest in that the ratio of the products varies with pressure. The results indicate that compound (A) is the initial product and that it is formed in a high vibrational level of its ground state, from which it is either stabilized by collisional deactivation or isomerizes to give (B) in a high vibrational level of its ground state. Product (B) in turn is either deactivated by collision or decomposes to give compound (C).

6.2.12 *Diazo-compounds*

The photolytic decomposition of diazo-compounds is widely used in synthetic organic chemistry for generating carbene intermediates. Methylene (CH_2 :) is the simplest carbene and it is of particular importance in relation to preparative chemistry because of the facility with which it adds across olefinic double bonds, to give cyclopropyl derivatives, or inserts into C—H bonds to give methylated derivatives. Methylene has two electrons which are not involved in bonding and it can therefore exist in a singlet state CH_2 ($\uparrow\downarrow$) or in a triplet state CH_2 ($\uparrow\uparrow$). Both singlet and triplet forms are produced when diazomethane is photolyzed in the vapour phase. The primary step in the photolysis produces singlet methylene which then converts to triplet methylene, the conversion being favoured at high inert gas pressures:

$$CH_2N_2 \xrightarrow{h\nu} CH_2(\uparrow\downarrow) + N_2 \xrightarrow[\text{gas}]{\text{inert}} \cdot CH_2(\uparrow\uparrow) + N_2$$

Triplet methylene can be prepared in solution without the intermediacy of the singlet form by photolyzing benzophenone in the presence of diazomethane. The mechanism of the reaction is as follows:

$$(C_6H_5)CO \xrightarrow{h\nu} (C_6H_5)_2CO(S_1)$$

$$(C_6H_5)CO\,(S_1) \longrightarrow (C_6H_5)_2CO(T_1)$$

$$(C_6H_5)CO\,(T_1) + CH_2N_2 \longrightarrow (C_6H_5)_2CO + CH_2(\uparrow\uparrow) + N_2$$

PHOTOCHEMICAL REACTIONS – II

Singlet and triplet methylene are of interest in that they react differently with olefinic systems. Singlet methylene adds across an olefinic bond in a single step retaining the stereo-arrangement of the groups in the olefin, whereas triplet methylene adds across an olefinic bond in a two-stage process and the addition is non-stereospecific. This difference in reaction mode is illustrated below for the reaction of singlet and triplet methylene with cis-but-2-ene:

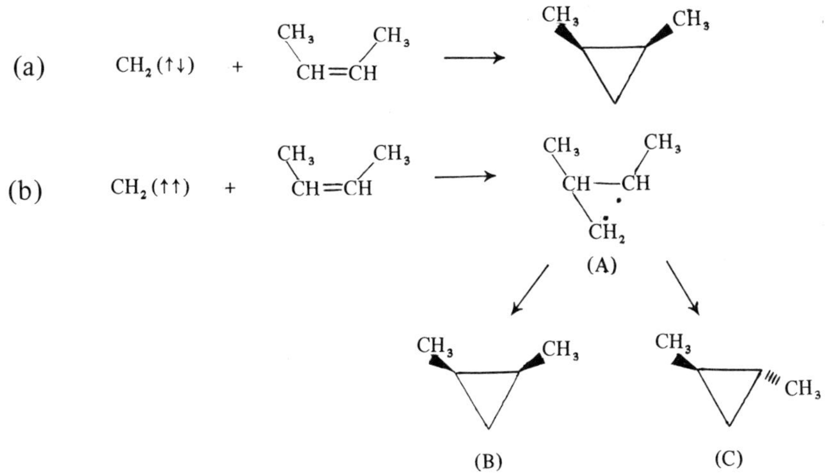

(a) $CH_2(\uparrow\downarrow)$ + cis-CH_3-CH=CH-CH_3 → cyclopropane derivative

(b) $CH_2(\uparrow\uparrow)$ + cis-CH_3-CH=CH-CH_3 → biradical (A) → (B) and (C)

In the reaction above with triplet methylene a biradical intermediate (A) is formed which can either ring close directly with the formation of the cyclopropane derivative (B) or rotate about the central C—C bond before ring closure, in which case the cyclopropane derivative (C) is formed. The original stereo-arrangment is retained in (B) but not in (C).

Carbenes other than methylene can be generated by photolysis of the appropriate parent compound. Although the carbenes so formed are unstable and cannot be isolated their formation is often clearly indicated by the structure of the products formed in the reaction, e.g.

(a) $(C_6H_5)_2CN_2 \xrightarrow{h\nu} (C_6H_5)_2C: + N_2$
$\searrow (C_6H_5)_2CN_2$
$(C_6H_5)_2C=N-N=C(C_6H_5)_2$

137

(b)

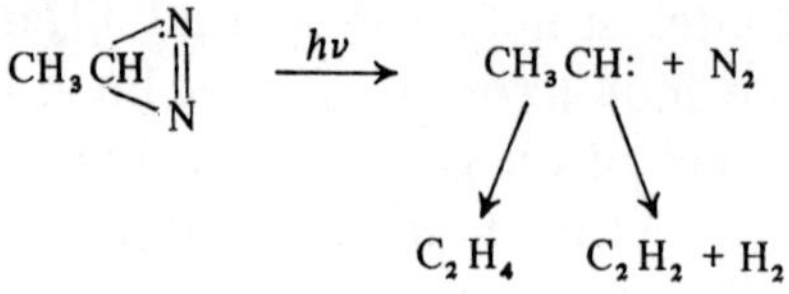

$$CH_3CH(N_2) \xrightarrow{h\nu} CH_3CH: + N_2$$

$$\swarrow \qquad \searrow$$

$$C_2H_4 \qquad C_2H_2 + H_2$$

6.2.13 Azides

Aliphatic azides decompose on irradiation to give a nitrene and nitrogen in the primary dissociative step:

$$R-\overset{-}{N}=\overset{+}{N}=N \xrightarrow{h\nu} R-N: + N_2$$

The nitrene formed in the primary step can react in any one of the three ways illustrated below for n-pentyl nitrene viz. (1) isomerization to an imine, (2) hydrogen abstraction from hydrogen-containing solvents to give an amine, and (3) 1,5-hydrogen abstraction (if possible) followed by cyclization to produce a pyrollidine:

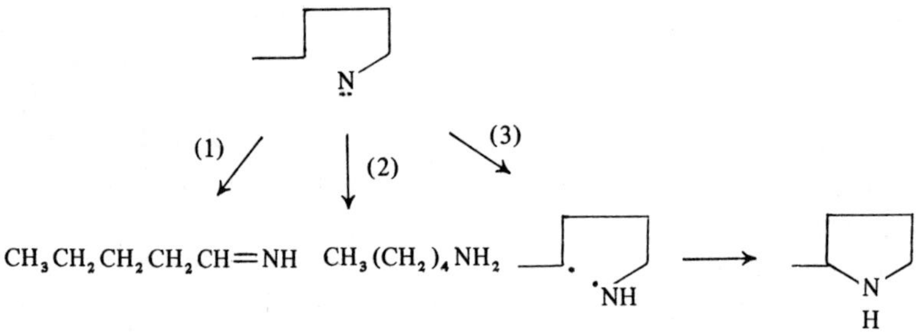

(1) $CH_3CH_2CH_2CH_2CH=NH$ (2) $CH_3(CH_2)_4NH_2$ (3) → pyrrolidine

Reactions of the type (1) and (2) shown above are observed to occur when cyclohexyl azide is photolyzed in cyclohexane:

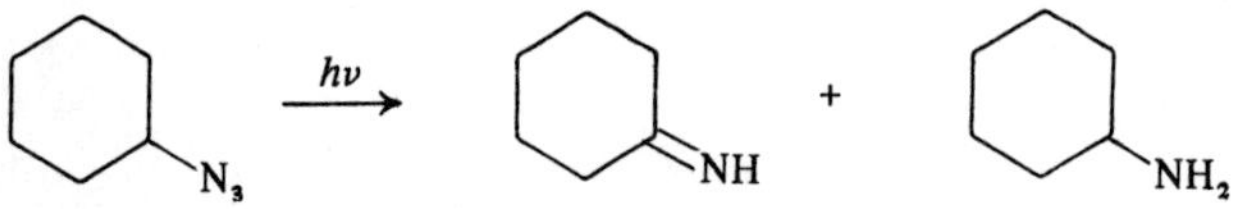

cyclohexyl-N_3 $\xrightarrow{h\nu}$ cyclohexyl=NH + cyclohexyl-NH_2

PHOTOCHEMICAL REACTIONS – II

Photolysis of unsaturated azides can differ from the photolysis of saturated azides in that the intermediate nitrene may isomerize to a cyano compound, e.g.

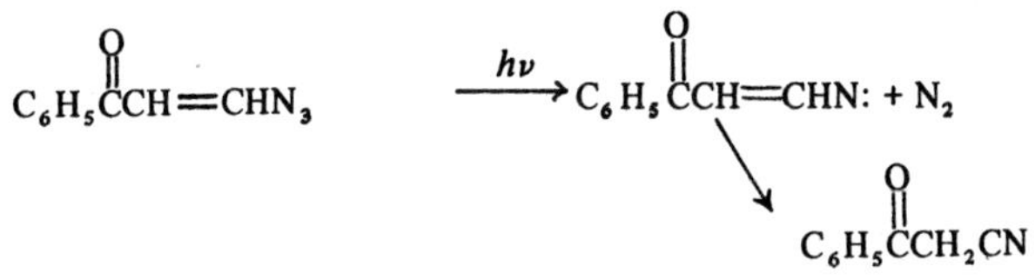

$$C_6H_5\overset{O}{\overset{\|}{C}}CH=CHN_3 \xrightarrow{h\nu} C_6H_5\overset{O}{\overset{\|}{C}}CH=CHN: + N_2$$

$$\downarrow$$

$$C_6H_5\overset{O}{\overset{\|}{C}}CH_2CN$$

Bibliography

Textbooks

J. G. CALVERT and J. N. PITTS, Jr., *Photochemistry* (Wiley, New York, 1966).
R. O. KHAN, *Organic Photochemistry* (McGraw-Hill, New York, 1966).
N. J. TURRO, *Molecular Photochemistry* (Benjamin, New York, 1966).
R. B. CUNDALL and A. GILBERT *Photochemistry*, (Nelson, London, 1970).
R. P. WAYNE, *Photochemistry* (Butterworths, London, 1970).
J. P. SIMONS, *Photochemistry and Spectroscopy* (Wiley-Interscience, London, 1971).
A. COX and T. J. KEMP, *Introductory Photochemistry* (McGraw-Hill, Maidenhead, 1971).

Review Texts

O. L. CHAPMAN, (Ed.), *Organic Photochemistry,* Vol. 1 (Arnold, London, 1967).
O. L. CHAPMAN, (Ed.), *Organic Photochemistry,* Vol. 2 (Dekker, New York, 1969).
J. N. PITTS, Jr., G. S. HAMMOND and W. A. NOYES, Jr., (Eds.), *Advances in Photochemistry,* Vols. 1–8 (Wiley-Interscience, New York, 1963–1971).
N. J. TURRO, *et al., Annual Survey of Photochemistry,* Vols. 1 and 2 (Wiley-Interscience, New York, 1969 and 1970).
D. BRYCE-SMITH, (Snr. Reporter), *Photochemistry,* Vols. 1 and 2 (The Chemical Society, London, 1970 and 1971).

Index

Absorbance, 4
Absorption coefficient, 4
Absorption intensity, integrated, 23
Absorption spectrum, 4
   detmn. of radiative lifetimes, 65
   information on excited states, 47
Acenaphthylene, dimerization of, 94
Acetone, 89, 100, 133
Acetophenone, 89
   addition to olefins, 100
   as sensitizer, 94
Acids, photolysis of, 127
Acrylonitrile, 102
Acyclic aliphatic ketones, photocyclization in, 111
Addition, photo-, 97–105
Alcohols, photolysis of, 128
Aldehydes, photolysis of, 124
Alkenyl bromides, preparation of, 121
Alkoxy radicals, 130–131
Alkyl aldehydes, addition to olefins, 100
Alkyl benzenes, photolysis of, 123
Alkyl bromides, preparation of, 121
Alkyl halides, photolysis of, 121
Alkyl iodides, photolysis of, 121
Alkyl ketones:
   addition to olefins, 100
   hydrogen abstraction from amines, 99–100
Allowed transition, 23–24
Amides, photolysis of, 132

Amines:
   addition reactions, 104, 134
   hydrogen abstraction from, 92, 99
Anilides, photolysis of, 133
Aniline, 134
Anthracene:
   dimerization of, 93–94
   9-methyl-, 94
   oxidation of, 107
9-Anthraldehyde, 94
Anthraquinone:
   absorption spectrum, 10
   as sensitizer, 95
Antibonding orbitals, 14–15, 26
Aromatic carbonyls:
   absorption spectra, 22
   addition to, 99
Aromatic halides, photolysis of, 122
Aromatic hydrocarbons:
   dimerization of, 94
   photolysis of, 123
Azides, photolysis of, 138
Azobisisobutyronitrile, 135
Azo-compounds, photolysis of, 134
Azoisopropane, 134

Band intensities, 22, 28, 47
Barton reaction, 132
Beer–Lambert law, 4
Beer law, 4

141

INDEX

Benzene:
   absorption spectrum, 5, 17, 114
   addition reactions, 102–104
   molecular orbitals in, 16
   photolysis of, 114, 123
Benzhydrol, 89, 92
Benzophenone:
   absorption spectrum, 5, 7
   addition reactions, 98, 100
   alkyl-, 90
   p-amino-, 90
   as sensitizer, 106, 133, 136
   o-diiodo-, 122
   energy transfer from, 53, 78, 95–96
   excited state properties, 44–46
   hydrogen abstraction by, 49, 88–89, 91–92
   o-methyl-, 90
   p-phenyl-, 89
   o-tertbutyl-, 91
Benzoquinone, 92, 98, 102
Benzpinacol, 89
Benzvalene, 114
Benzyne, 122
Biacetyl:
   energy transfer from, 52, 75
   energy transfer to, 75, 79, 81
Bicyclo [4,2,0] octa-2,4-dienes, 124
Biradical intermediates, 98–99, 124–125, 135, 137
Bond dissociation, 116–119
Bonding orbitals, 14–15, 26
Bromine, 120
Bromotrichloromethane, 122
Butadiene, dimerization of, 94
But-2-ene, 137
n-Butyric acid, 128

Cage effect, 119
Carbenes, 136–138
Carbon dioxide, elimination of, 127

Carbon monoxide, elimination of, 124, 125, 127
Carbon tetrabromide, 122
Carbon tetrachloride, 122
Carbonyl compounds, α,β-unsaturated, 96–97
Carbonyl group:
   addition to, 97–102
   electronic configuration, 20
   electron distribution in, 89
   electronic transitions in, 20–21
   hydrogen abstraction by, 89–93
   molecular orbitals of, 18
Chain reaction, 9, 120
Charge-transfer complex, in photoreduction, 92
Charge-transfer state, 90
Chlorine, 120
1-Chloronaphthalene, 26
Collisional deactivation, 49, 119
Coumarin, 96
Coupled transitions, 55
Critical distance, 57
Cyclization, 110
Cyclobutanol formation, 111, 127
Cyclohexenone, 97
Cyclohexyl azide, 138
Cyclo-octadiene, 98, 110
Cyclopentadiene:
   addition to, 101
   dimerization of, 95
Cyclopentanone, 124

Debye equation, 50
Decarbonylation, 124–126
Decarboxylation, 128
Delayed fluorescence, 38
Dewar benzene, 114
Dianthracene, 94
Diatomic molecules:
   absorption spectra, 29–30
   potential energy curves, 27–28, 118
   predissociation, 118

INDEX

Diazo-compounds, photolysis of, 136
Diazomethane, 136
Dibenzyl peroxide, 131
3,4-Di-*t*-butylcyclobutanedione, 126
Di-*t*-butylcyclopropanone, 126
Di-*t*-butylperoxide, 130
1,2-Dicyanoethylene, 100
Diels-Alder adduct, 91, 122
1,2-Diketones, 112, 126
Dimerization processes, 93–97
Dimethylacetylene dicarboxylate, 102
2,5-Dimethylfuran, 107
Diphenylacetylene, 102
4,4-Diphenylcyclohexa-2,5-diene-1-one, 111
1,1-Diphenylethylene, 108
2,3-Diphenylphenol, 111
1,2-Diphenylpropene, 108
Discharge lamps, 9
Dissociation, 116–119

Einstein of radiation, 7
Electromagnetic spectrum, 2–3
Electron donation, 92
Electronic states, 19
Electronic transitions:
    identification of, 21, 48
    types of, 15–17
Elimination reactions, 123–125, 135
Emission spectra, 30
Energy, frequency and wavelength relationship, 2
Energy transfer:
    intramolecular, 54
    involving oxygen, 106
    kinetics of, 71–82
    long-range, 55–58
    radiative, 50–51
    short-range, 51–55
    singlet–singlet, 52–54, 57, 78–82
    triplet–singlet, 57
    triplet–triplet, 52–54, 57, 72–78, 96, 99–101

Esters, photolysis of, 127
Ethers, photolysis of, 128
Ethyl iodide, 26
Excimer, 38
Excitation energy, 3, 7,
Excited state lifetimes, 60
Excited state molecule, 2

Filters, 11
Flash photolysis, 62–64, 76–78
Fluoranthene, 76
9-Fluorenone, 92
Fluorescein, 107
Fluorescence, 30–33, 36–43
    delayed, 38
    emission spectra, 31, 33, 41
    excimer, 38
    kinetics of, 65
    lifetime measurement, 61, 64
    prompt, 38
    quantum yield values, 43, 44, 48
Forbidden transition, 23–24
Formamide, 133
Franck-Condon principle, 28, 33
Free radical polymerization, 131
Frequency, 2
Fries rearrangement, 112
Fulvene, 114
Fumaric acid, 108
Furan, 125

Gerade wave function, 26
Ground state configuration, 20
Grotthus-Draper principle, 3

Halogenation, photo, 120
Halogens, 116
    bond dissociation energies, 116
    photolysis of, 120
Heavy atom effect, 25, 43
    of 1-chloronaphthalene, 26
    of ethyl iodide, 26

143

INDEX

Hot radicals, 119
Hydrocarbons, photolysis of, 123
Hydrogen abstraction, 88–93, 111
Hydrogen bromide, addition to olefins and acetylenes, 120–121
Hydrogen elimination, 123
Hydrogen halides:
    elimination of, 121
    photolysis of, 120
Hydroxy-hydroperoxide formation, 106

Inner electrons, 20
Integrated absorption intensity, 23
Intermolecular processes:
    deactivation by, 48–50
    kinetics of, 71–87
Internal conversion, 34–36, 70
Internuclear separation, 28, 117–118
Intersystem crossing, 34–36
    kinetics of, 66–70, 78
    quantum yields of, 41, 44, 54
Intramolecular processes:
    cyclization, 110
    deactivation by, 34–48
    energy transfer, 54
    hydrogen abstraction, 88–93, 127, 131
    kinetics of, 59–70
Isomerization:
    cis-trans, 107–109
    structural, 107, 113
    valence bond, 107, 110
Isoprene, dimerization of, 94
Isopropanol, 130
    hydrogen abstraction from, 89, 92
    sensitized oxidation of, 106

Keto-ethers, 112
Ketones:
    addition reactions of, 97–98, 101
    cyclization of, 111
    photofragmentation of, 124
    photoreduction of, 89

Lambert law, 4
Lamps, mercury, 9
Lasers, in flash photolysis, 63
Lifetimes of excited states, 37, 48, 60
    singlet states, 64, 66
    triplet states, 64, 66
Liquid phase photolysis, 119

Maleic anhydride, 91, 97
    addition reactions of, 104–105, 129
Maleimide, 103
Mercury discharge lamps, 9
1-Methoxy-1-butene, 101
2-Methyl-but-2-ene, 98
Methylene, 136–137
3-Methylene-1,5-hexadiene, 110
Molecular orbitals, 14
    energies of, 18
    HOMO, 19
    LVMO, 19
Monochromatic radiation, 4
Multiplicity of states, 19

N,N-Dimethylacetamide, 92
n-orbitals, 17
(n, π^*) states, properties of, 44–48, 89
1-Naphthaldehyde, 89
Naphthalene:
    absorption spectrum, 10
    energy levels of, 52
    energy transfer from, 79
    energy transfer to, 53, 76, 78
    excited state properties, 44
1,4-Naphthaquinone, 97
1-Naphthyl acetic acid, 128
2-Naphthylphenylketone, 26
Natural radiative lifetime, 37, 64–65
Nitrenes, 138
Nitrites, photolysis of, 131
Non-bonding orbitals, 17
Norbornadiene, 110
Norbornene, 100
Norrish Type I process, 124
Norrish Type II process, 126

INDEX

Observed radiative lifetime, 37, 64
2,4,6-Octatriene, 110
Optical dissociation, 111
Orbital symmetry, 25
Organic halides, photolysis of, 121
Oscillator strength, 23–24
Oxetan formation, 97–102
Oxeten intermediate, 101
Oxidation, photo, 105–106
Oxygen:
    addition reactions of, 105–107
    enhancement of singlet–triplet transitions, 25
    quencher of triplet states, 50, 80, 103

π-orbitals, 15
(π, π^*) states, properties of, 44–48, 89
Paramagnetic effect, 25
Parity, 26
Pauli exclusion principle, 19
1,3-Pentadiene, 108
t-Pentyl nitrite, 131
Peroxides, photolysis of, 130
Perpendicular triplet state, 108
Phenanthrene, 105
Phosphorescence, 30–33, 36–43
    emission spectrum, 31
    kinetics of, 65
    lifetime measurement, 61
    quantum yield values, 41, 43, 44, 48
Photo-addition:
    to aromatic compounds, 102
    to carbonyl compounds, 97
Photoamidation, 133
Photochemical reactions, 88–139
    kinetics of, 82–87
Photochemical reactors, 11
Photocyclization, 110
Photodimerization:
    of aromatic hydrocarbons, 94
    of conjugated dienes, 94–96
    of α,β unsaturated carbonyls, 96–97

Photofragementation, 116–119
    of acids, 127
    of alcohols, 128
    of aldehydes, 124
    of amides, 132
    of amines, 133
    of azides, 138
    of azo-compounds, 134
    of esters, 127
    of ethers, 128
    of ketones, 124
    of diazo-compounds, 136
    of halogens and hydrogen halides, 120
    of hydrocarbons, 123
    of ketones, 124
    of nitrites, 131
    of organic halides, 121
    of peroxides, 130
Photohalogenation, 120
Photo-oxidation:
    of alcohols, 106
    of conjugated systems, 106
    Type I mechanism, 105
    Type II mechanism, 106
Photorearrangement:
    cis-trans isomerization, 107
    Fries rearrangement, 112
    intramolecular cyclization, 110
    of benzene, 114
Photoreduction, of carbonyl compounds, 88–93
Planck's constant, 2
Polyatomic molecules:
    photodissociation of, 119
    potential energy functions of, 27, 118
Polyhalomethanes, photolysis of, 122
Potential energy curves, 28–29
Predissociation, 118
Prompt fluorescence, 38
n-Propanol, 129
Propionaldehyde, 98
Pyrene, 38, 52

INDEX

Pyrrole, 104
Pyruvic acid, 128

Quantum number:
   rotational, 6–7
   spin, 19
   vibrational, 6
Quantum yield, 8, 39
Quenching, 72
   by oxygen, 50, 80, 103

Radiation, 2
   energy of, 8, 116
Radiation filters, 11–12
Radiationless transitions, 34–36, 39
Radiative lifetimes, 37, 41, 64–65
Radiative transitions, 30, 36–43
Radical formation, 116, 119, 122, 130, 135
Rate constant, 36, 40, 59
   diffusion controlled, 50
   for dimerization of thymine, 86
   for fluorescence, 60, 66
   for internal conversion, 60, 70
   for intersystem crossing, 60, 70
   for phosphorescence, 60, 66
   for long-range energy transfer, 57
   for singlet–singlet energy transfer, 81–82
   for triplet–triplet energy transfer, 75, 78
Rearrangement reactions, 107–115
Rose Bengal, 107
Rotational energy levels, 7
Rotational quantum number, 6–7

σ-orbitals, 15
Selection rules, 24
   electron spin, 25
   momentum, 27
   orbital symmetry, 25
   parity, 26
Self-quenching, 49–50

Singlet–singlet energy transfer, 52–54, 57, 78–82
Singlet state, 19
   identification of type, 47–48
Singlet–triplet transitions, 25–26
   enhancement of, 25–26
Solvent:
   cage effect, 119
   effect on band positions, 21
   effect on excited state energy, 32
Spin quantum numbers, 19
Stark-Einstein principle, 3
Stationary state hypothesis, 67
Stern-Volmer equation and plots, 74
Stilbenes, 108

Tetrahydrofuran, 129
Tetrahydropyran, 130
2,2,4,4,-Tetramethylcyclobutane-1,3-dione, 125
Thymine, 87
Tributylstannane, 90
Trifluoroiodomethane, 122
Triple–singlet energy transfer, 57
Triplet state, 19
   identification of type, 47–48
   perpendicular, 108
Triplet–triplet absorption spectra, 63
Triplet–triplet energy transfer, 52–54, 57, 72–78, 96, 99–101

Ungerade wave function, 26
α,β-unsaturated carbonyls, dimerization of, 96–97

Valerophenones, 91
Vapour phase photolysis, 119
Vibrational energy levels, 6, 27, 117
Vibrational quantum number, 6

Wave length, 2
Wave number, 2
Wigner spin conservation rule, 51